T0230436

IUTAM

Peter Eberhard · Stephen Juhasz
Editors

# IUTAM

## A Short History

Second Edition

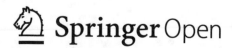 Springer Open

*Editors*
Peter Eberhard
Institute of Engineering and Computational
    Mechanics
University of Stuttgart
Stuttgart
Germany

Stephen Juhasz (26.12.1913–19.6.2013)
San Antonio, TX
USA

ISBN 978-3-319-80953-3          ISBN 978-3-319-31063-3    (eBook)
DOI 10.1007/978-3-319-31063-3

# Contents

*This short history is dedicated to the memory
of Professor Theodore von Karman
1881–1963 who had an essential role
in the formation of IUTAM*

Aachen, Germany in the early 1920's

*Picture Source*: *The Wild and Beyond: Theodore von Karman, Pioneer in Aviation and Pathfinder in Space* by Theodore von Karman, with Lee Edson. Little Brown & Co. 1967
Courtesy of the Archives, California Institute of Technology, depository of the von Karman collections

**Disclaimer**

*The statements and opinions expressed in this volume are those of the authors and not necessarily of the International Union of Theoretical and Applied Mechanics (IUTAM), the organizations of the contributors, or the publisher.*

**Picture Credits**

*Abramson, H.N.*
*Batchelor, G.K.*
*Caltech Archives*
*Crandall, S.H.*
*Eberhard, P.*
*Godstein, J.*
*Hoff, J.*
*Hult, J.*
*Igenbergs, E.B.*
*Janssens, P.*
*Juhasz, S.*
*Lippmann, H.*
*MIT Archives*
*Mikhailov, G.K.*
*Olhoff, N.*
*Phillips, J.*
*Rimrott, F.P.J.*
*Schiehlen, W.*
*Tabarrok, B.*
*van Campen-Stuurman, A.*
*and official photographers*

# Preface

Since the often so-called 0th Congress in Innsbruck there have been many mechanics-related activities worldwide and it was a wise consequence to found in 1946 the International Union of Theoretical and Applied Mechanics, abbreviated IUTAM. In 2016 we can celebrate in Montreal, Canada, the 70th anniversary of IUTAM at the 24th International Congress on Theoretical and Applied Mechanics, abbreviated ICTAM. Besides these World Congresses, by the end of 2015 more than 380 symposia and summer schools have been organized by IUTAM.

Looking through the names of researchers involved in IUTAM, either as Officer, Bureau Member, General Assembly Member, Congress Committee Member, Symposia Panel Member, Symposia Organizer, or in another function for IUTAM, reads like a who-is-who in mechanics. So many famous scientists of the past have been actively involved that IUTAM can be truly proud to play such a prominent part in the history of mechanics. Also among the researchers currently active there are many famous names from all over the world. These people organize conferences, are editors of scientific journals, serve on all kinds of boards, lead big research programs, and supervise tens of thousands of doctoral students among various other contributions to our community. Research organized in IUTAM serves the promotion of science and the advancement of human society very actively and contributes to many urgent challenges and topics.

In 1988, the first edition of this book was published which covered the period until 1984. Stephen Juhasz worked for a long time to accurately collect all kinds of information related to the history of IUTAM and prepared a manuscript to report with many details about the creation and development of IUTAM. This book was published by Springer and turned out to be a most valuable reference. The basic structure of the original book was such that at first, several chapters related to IUTAM's history or certain organizational aspects were presented. Then, an extensive data and statistics part was included and, finally, photos from the history of IUTAM were shown.

Eminent scientists from mechanics contributed to the book and its structure proved to be useful. Colleagues who participated in the history of IUTAM found the activities of themselves and their friends, but also the general audience saw clearly why mechanics plays such an important role in universities and engineering education all over the world.

When the first edition of this book became available, I had just entered the university. However, during my study times and later as a research assistant, IUTAM was showing up to me everywhere. I attended IUTAM symposia and ICTAM World Congresses and especially my close cooperation with Werner Schiehlen introduced me early to IUTAM's work and many of its key figures. Later, when I became Treasurer of IUTAM, I became interested in IUTAM's history but found it a pity that Stephen Juhasz' great collection of historical data became a bit outdated 30 years after its presentation to the public. The somehow naive idea came up, to create a new edition by just updating a few tables. This turned out to be not appropriate and a lot of intense work was required to prepare the now completed book.

The basic concept was kept. However, since the time J. Lighthill wrote his chapter on "What is Mechanics?" many new developments had occurred and so T. Pedley and V. Tvergaard contributed two new chapters on the topics that gained importance during the past decades. The description of ICTAMS's early history by G. Battimelli was left largely unchanged and also the description of IUTAM's history up to the mid-1980s by J. Hult and N. Hoff remained nearly unchanged. IUTAM's development since then was summarized by W. Schiehlen and it is fascinating to see that many good things of the past were preserved while IUTAM adjusted to modern requests and changed many things in the past decades. These existing chapters were just slightly revised and corrected and the other parts were fitted to the style of the complete book. Some figures were added or moved from the appendices to the main part of the book.

The chapters on the Congresses and Symposia originally written by S. Juhasz and also the parts about the General Assemblies, originally written by D. Drucker and S. Juhasz, and the Reports by F. Niordson and S. Juhasz, required major changes and updates to reflect the current state of our union.

All the tables and statistics have been reorganized, updated, and completed. This required a lot of efforts and many iterations and corrections. The collected IUTAM annual reports were the most important sources of information, but also a lot of library work was needed. In total, 22 appendices with more than 100 pages document now in great detail most aspects of IUTAM's history and especially honor all the current and past colleagues who contributed so much to the visibility of IUTAM.

Finally, the photo part was revised and extended. More than 30 pages now show photos from all epochs of IUTAM's history and most of its key players. While many things changed, it is visible that the friendship, the international attitude, and the sense of serving a fascinating community remained the same from the very beginning to today.

Of course, such a book project is not the work of a single person alone. It is very sad that Stephen Juhasz passed away in 2013 at the age of 99. He dedicated so much time and effort to the first edition of this book and I am still highly impressed. I deeply regret that we never met in person and that I was never able to talk with him about this project. However, I am confident that he would have been delighted about seeing the revised and updated version of his book about IUTAM's history.

I am fortunate to enjoy the friendship of Werner Schiehlen for more than 25 years. Already when I was his doctoral student, a deep personal relation developed and it is our mutual pleasure to see each other every week. Werner Schiehlen provided besides a new chapter also several photos and a lot of information and advice. Also, he did a very careful proofreading of the manuscript as did Carl Herakovich, Frederic Dias, and Dick van Campen.

I am indebted to all the contributors of chapters for the first and the new edition. Except G. Battimelli, sadly all contributors of the original chapters have passed away and I hope to have made the editorial changes and revisions in their spirit. In total, seven former or current presidents of IUTAM as well as some other famous scientists contributed to this book, and so it is my hope that it provides an appropriate framework for the description of IUTAM's history.

Aatje van Campen-Stuurman and Jim Phillips provided additional photos. Some further photos were taken from the ICTAM reports and web pages.

Several students of mine, Sibylle Berger and Annelie Schädt, as well as my former doctoral student Dr.-Ing. Trong Phu Do spent many weeks collecting information and updating tables and my doctoral student Dipl.-Ing. Philipp Wahl rearranged all photos. All these persons deserve my thanks and appreciation.

I also want to thank the IUTAM Bureau for providing some support to make it financially possible to provide this book free of any cost as downloadable Open-Access publication and at the same time as a printed book on paper. I am also indebted to Nathalie Jacobs, our editor at Springer. She strongly motivated me to start this project, provided scanned files of the first edition, and also provided very fair conditions to IUTAM. I know Nathalie for many years and from several joint projects and it is always a great pleasure to cooperate with her and Springer and to experience her enthusiasm and great professional abilities.

Despite all efforts and all proofreading there are certainly still some mistakes remaining in this book. Please send me all corrections by email to peter.eberhard@itm.uni-stuttgart.de or by regular mail. An erratum or updates will be provided at www.itm.uni-stuttgart.de/iutam-book-2016 and every advice and all comments are highly welcome.

It is the hope of all authors that this book might be a valuable source of information for everybody interested in the history of mechanics and especially of IUTAM. It is a characteristic of IUTAM that it preserves important traditions while at the same time it adjusts to new challenges and modernizes its structures and processes. IUTAM's history will go on and it will be highly interesting to see what the future will bring. We are deeply convinced that mechanics will always be one of the most important foundations of engineering while also being at the forefront of advanced developments. Hopefully, in the future somebody will accept the duty to update again this book and I am looking forward to reading then about all developments which are by now still far in the future.

Leonberg, Germany                                              Peter Eberhard
2015

*Photo*: Book presentation by S. Juhasz (Grenoble 1988)

# What Is Mechanics?

**Sir James Lighthill**

The International Union of Theoretical and Applied Mechanics is active both in Theoretical Mechanics and in Applied Mechanics. Beyond these, however, it is concerned with that fruitful interaction between them which, increasingly, has constituted one of the great scientific success stories of the past 300 years.

## A. Theoretical Framework

### 1. Newtonian Foundations

Mechanics was the first science for which a systematic theoretical framework founded on mathematics was created, as pioneered in Newton's 'Principia' (first published in 1687). Although many already existing ideas were incorporated in 'Principia', it represented a fundamentally new overall conception of how a major science can be characterized in terms of laws which in form are of an explicitly mathematical character.

The underlying strategy developed by Newton for this purpose, and since carried much further by his successors, was a 'divide and rule' strategy. It proved equally appropriate to either of the two major parts of mechanics; that is, to statics, or dynamics; concerned with how motion is prevented, or governed, respectively.

S.J. Lighthill (1924–1998)
University College London, London, UK

© The Author(s) 2016
P. Eberhard and S. Juhasz (eds.), *IUTAM*,
DOI 10.1007/978-3-319-31063-3_1

1

Thus, Newton's strategy requires first that we divide up or classify different types of *force*; representing different influences which in combination may act to prevent motion, or to govern it when it occurs. *Matter*, again, is to be thought of as divided up into particles of relatively small size, which are subject to forces of two main kinds: *external* forces (such as gravity, for example), along with those internal forces which act between a pair of particles and which necessarily assume equal and opposite values.

Motion may be prevented in a system where the sum (in a vectorial sense) of all the different forces acting on each and every particle of the system is zero. Wherever that sum is nonzero, however, its value specifies the rate of change of the *momentum* (mass times velocity) for the particle's motion; which, therefore, it governs.

Besides the twofold application of the 'Divide and Rule' strategy within statics (through classification of forces, and through the subdivision of matter into particles), dynamics necessitates a third subdivision of time into brief instants. During each such instant a particle's momentum changes by an amount equal to the instant's duration times the sum of the forces acting on the particle.

## 2. Mathematical Formulation of Newtonian Mechanics

For the dynamics of a body small enough (relative to an entire system) to be treated as a single particle, this principle can be expressed mathematically as an *ordinary differential equation* with time as the independent variable. Motions of planets and of their satellites, for example, lend themselves to interpretation in terms of solutions of such ordinary differential equations.

In other circumstances, when matter is viewed as subdivided into many particles, even the laws of statics may need to be expressed as *partial* differential equations with space coordinates as independent variables. Their form depends upon an assumed knowledge of *material properties*, including knowledge of how internal stresses (the internal forces acting between adjacent particles, divided by their area of contact) depend upon other factors including the body's degree of deformation. Such a partial differential equation may express the fact that the sum of all forces on a particle, including external forces as well as the internal forces of interaction with adjacent particles, is zero.

Dynamics, on the other hand, allows that sum to take a nonzero value which determines how the momentum of the particle concerned changes from one brief instant to the next. Analytical expression of results such as this takes the form of partial differential equations with space coordinates *and also the time* as independent variables. Modern treatments of such problems using high-speed computers may approximate the partial differential equations by means of finite differences; or may, alternatively, apply the basic principles of mechanics directly to carefully selected particles of matter known as *finite elements*.

Theoretical mechanics has continued over the years to make massive progress through (i) improved representation of material properties, (ii) incorporation of additional types of external force, and (iii) advances in analytical or numerical methods. Brief indications of these main lines of progress will now be given.

## 3. Improved Representation of Material Properties

For matter in the solid state a rather early initiation, and subsequent extended development, of a comprehensive theory of elasticity, allowing stress to depend linearly on the measure of deformation known as strain, led to impressive results in both statics and dynamics. It was then followed successively by a range of more refined theories (including, for example, the theory of plasticity) which especially need to be applied in cases of *large deformation*. Important newer developments have resulted from the treatment of 'materials with memory'.

For matter in either of the two fluid states (liquid or gas), no serious problems are posed by statics; mainly, because stresses in a fluid at rest take the form of a simple pressure acting equally in all directions, as explained in elementary hydrostatics texts. The dynamics of fluids, however, is affected by additional stresses associated for common fluids with their viscosity: and such stresses prove to be important even for fluids of very small viscosity. This is because the condition satisfied by a fluid motion at a solid boundary requires its velocity to vary steeply across a thin boundary layer, where viscous stresses are substantial and may significantly affect whether or not the *boundary layer*, as a result of separating prematurely from such a solid boundary, changes drastically the character of the whole flow.

Different relationships of stress to motion, more complicated than can be described by a simple viscosity coefficient, govern the flow properties of certain substances. Analysis of these relationships and their consequences forms the branch of mechanics known as rheology.

Other material properties which may be important include those familiar from thermodynamics. They are concerned with the convertibility of energy between its various forms, and with 'equations of state'. In a fluid, for example, these relate changes in the fluid density (mass per unit volume) to changes in pressure and also temperature. For gases, furthermore, kinetic theory (applying mechanical laws to the molecules themselves) may give valuable information about material properties which can critically affect gas dynamics, including not only viscous action but also *heat transfer* (which, besides influencing dynamical behaviour, is also of importance in its own right).

# 4. Consideration of Additional External Forces

External forces which in mechanics may need to be considered together with gravity include *electromagnetic* forces; whose effect can be substantial, especially for good conductors. These include both solid and liquid metals, and also ionized gases, some of whose electromagnetic properties are found to deserve study by means of kinetic theory.

Systems subjected to a general rotation may be analyzed by means of space coordinates related to a rotating frame of reference. Dynamics in such a rotating frame effectively incorporates additional forces (the centrifugal and Coriolis forces) which influence greatly the behaviour of technically important *gyroscopic* systems, and also the dynamics of the ocean or the atmosphere.

# 5. Advances in Mathematical Methods

Beginning with the analytical mechanics of Lagrange, mathematical theories founded upon energy considerations have proved to be of very general and wide-spread applicability. Some special mathematical methods of importance to theoretical mechanics include linear and nonlinear theories of *vibrations and waves*. These, for example, have proved to be of great value for acoustics by elucidating not only the types of vibrations which our ears detect but also their transmission to our ears as waves travelling through solid or fluid media.

Theories of *stability* are also most important. For statics, they can distinguish between an equilibrium which can persist and one which will collapse in response to a small disturbance of a particular shape (as when a solid shell buckles under load). For dynamics, they can distinguish between steady motions able to persist and those which will break up through instability into either regular or randomised disturbances. Description of instability, phenomena in terms of the generic process known as 'bifurcation' have proved increasingly fruitful.

More recently, general studies of *dynamical systems* have given the broad analytical background to two important tendencies: the tendency for certain continuously varying motions to develop discontinuities ('catastrophes', as when an acoustic wave develops into a shock wave); and the tendency for certain well ordered motions to develop into random motions ('chaos'; a process long studied in the dynamics of fluids as the tendency for laminar flows to develop into turbulence, but now recognized as relevant also to many solid systems with only a few degrees of freedom). Much of the previously existing knowledge and understanding in these areas (for example, of shock waves or turbulence) has been valuably extended and systematized by these new theories.

In all periods, many important mathematical advances have been initiated by research workers in mechanics. Thus the science of mathematics itself was a major beneficiary from progress in theoretical mechanics.

# B. Examples of the Applications of Mechanics

## 1. Applied Mechanics and its Growing Utilisation of Theoretical Mechanics

The applications of mechanics are found in many scientific fields, some of which (like astronomy, oceanography, and meteorology) have already been referred to; as well as in most of the principal subdivisions of engineering and technology. Into each particular scientific or technological field the process of penetration of any refined ideas from theoretical mechanics has been slow, for a good reason: the scientific phenomena needing to be elucidated, or the technological objectives needing to be met, were in most cases much too complicated for effective treatment by the methods of theoretical mechanics during their early stages of development.

In these circumstances, practitioners of the disciplines concerned needed to concentrate above all on devising ingenious *measurement techniques* appropriate to a particular area of application, and on amassing useful empirical rules for correlating data obtained with these techniques. Such rules might be expressed in phraseology using some of the simpler ideas from theoretical mechanics, but they were not based on detailed theoretical analysis.

Great achievements were to result from applied mechanics development using these empirical methods. In each area the approach in question has, furthermore, continued to make good progress, especially during periods immediately following the introduction of a new measurement technique.

At the same time the general progress in theoretical mechanics has successfully taken place, partly under its own momentum and partly under the stimulus of challenges posed by the complicated problems needing to be solved in particular areas of application. Gradually, within each field, theoretical analysis of 'model problems' has become recognized as making a truly valuable contribution to studies of the complicated 'real problems' which they were designed to model. Even when the agreement between experiment and theory was not very excellent such analysis might permit a most useful extrapolation of the available experimental data to other conditions for which additional experiments would be too difficult or costly or time-consuming. In time, the models became more and more comprehensive, and correspondingly more valuable.

These continuing processes, in which theoretical mechanics both makes valued contributions to, and is stimulated by challenges derived from, applied mechanics, have increasingly brought the two parts of mechanics much closer together in a fruitfully cooperative unity. Some examples of how this happened in particular areas of application are given in the rest of this note.

## 2. Structural Engineering

In structural engineering, early empirical developments had led to great achievements which included such refined designs as medieval cathedrals. Yet gradually methods became more analytical as it became possible to estimate loads and the *stress distributions* they would produce, first in single beams and then in frameworks; and to compare stresses so calculated with yield stresses characteristic of the materials used. *Stability* calculations were initiated too, by Euler, so as to yield estimates of critical loads for buckling.

Nevertheless, over long periods it was necessary to overdesign structures by 'factors of safety' as high as 10; that is, to design them so that calculated loads for yielding or buckling were ten times more than the estimated worst-case loads. This need arose from a great multiplicity of uncertainties; for example, in the field of stress analysis, or concerning the reliability of assumptions about material properties or load estimates.

Later, the study of frameworks and similar structures became much more thorough; it took increasingly into account the problems of stress concentrations at junctions between elements, and of detailed design to minimise these. Also, 'limit design' methods were developed (studying the behaviour of a framework after early applications of load have produced certain plastic deformations).

Furthermore, the inevitable presence in a structure of particular types of *imperfection* which adversely influence structural integrity was increasingly allowed for. These included geometrical imperfections powerfully affect buckling behaviour.

Again, small material imperfections including dislocations and cracks were intensely studied in relation to their capacity for growth leading to plastic deformation or to fracture. This, for example, allowed the phenomenon of metal *fatigue* under cyclic loading (first discovered empirically) to become clearly interpretable in terms of crack growth, and there were similar successes in the area of metal *creep*. In the meantime, the importance of composite materials (from concrete to carbon-fiber-reinforced plastics) became increasingly recognized, and this too stimulated many major new developments in theoretical mechanics.

The enormous economic advantage of structure-weight minimisation for the design of aircraft and spacecraft gave a special boost to refined structural analysis aimed at achieving accuracies such as would permit 'factors of safety' to be brought down from former figures like 10 to modern figures around 1.5 or even less. They also encouraged particular mathematical developments (for example, in the theory of shells) of very great subtlety.

At the same time, empirical methods advanced in parallel; particularly through the development of strain-gauge technology, of various optical techniques and of a wide range of convenient methods for loading simulation. The modern structural engineer uses an admirably well integrated blend of empirical activity (doing experiments and using compendious data from earlier experiments) with the analytic and computational methods of theoretical mechanics.

# 3. Hydraulics

The study of flow in man-made systems of pipes and channels goes back to the beginnings of history with the irrigation achievements in Egypt and Mesopotamia which made possible the first great civilisations. Then it developed further in response to successive needs, from (for example) the Roman Empire's large-scale aqueduct and hypocaust systems to the remarkable drainage and irrigation schemes carried out in the early years of the Dutch Republic.

Theoretical mechanics made its impact on the subject in the eighteenth century when Daniel Bernoulli showed how a steady stream, according to Newton's Laws, would in the absence of frictional forces conserve what we now call its *total head*. Much of the ensuing development of hydraulics (at first, mainly empirical) was expressed in terms of laws governing how much total head is frictionally dissipated by different types of steady stream under different conditions. Some puzzling features of these laws, which recognize (for example) how streams progressing along pipes lose total head far less steeply where the pipe cross-section contracts than where it expands, became intelligible in terms of theoretical mechanics only when the conditions governing *boundary layer separation* in a fluid flow had been elucidated both theoretically and empirically.

Other developments in the two-dimensional (and three-dimensional) modelling of fluid motions allowed improved design of pipe and channel junctions and detailed calculations of flows such as those in the neighbourhood of sluice-gates. More recently, turbulence modelling played a growing part in such work. In the meantime, the increasingly refined theory of *surface wave* phenomena had been most valuably applied so as to understand how stationary waves, standing waves and travelling waves are generated by flows in open channels. Also, *cavitation* theory elucidated effects due to the motion of a liquid generating locally negative pressures which cause the spontaneous appearance of bubbles.

In parallel with all the theoretical developments, laboratory studies utilising carefully instrumented flumes and water-tunnels and other hydraulic modelling devices have added greatly to knowledge in all parts of this discipline. Modern hydraulic design rests on calculations utilising a well integrated combination of results emanating both from theoretical mechanics and from experiments in the laboratory and in the field.

# 4. Mechanical Engineering

Building upon the long-established empirical technology of water-powered and wind-powered mill machinery, mechanical engineering began to develop rapidly after steam-power was introduced in the eighteenth century. Ideas from theoretical mechanics played an increasing role in these developments.

Thus, the part of dynamics known as kinematics (the *geometrical* description of relative motions) influenced the detailed design of trains of *gearing* mechanisms of all kinds; and, especially, the shaping of surfaces required to be in contact from gear-teeth to screws. The part known as kinetics (relation of motion to forces and to the *work* done by these) influenced the design not only of the actual means for transferring power but also of the devices for controlling the resulting motions. An increasingly important contribution was made by the theory of vibrations of mechanical systems, with its fundamental elucidation of the concept of normal modes of vibration each with its own natural frequency; and of the dominating influence of *resonance* (coincidence between forcing frequency and natural frequency) in generating these usually *unwanted* vibrations.

By the last quarter of the nineteenth century the mechanics of fluids had begun to exert a major influence on mechanical engineering design; as when, for example, a hydrodynamic theory of lubrication related the forces sustainable in a journal bearing to the geometry and viscous properties of the oil film. Above all, development of the *steam turbine* called for meticulous shaping of steam passages and of turbine blades, taking into account the interacting dynamics and thermodynamics of fast-moving steam. A new understanding of heat transfer, based on the fundamental mechanics of fluids, was at the same time being established.

Systems for efficient exchange of energy (in either direction) between streams of fluid and rotating blades were then progressively developed for the successive requirements of fans, propellers, fluid transmissions, compressors and gas turbines; with applications in aeronautics and many other fields. More and more, these developments rested upon refined representations of the dynamics of fluids and blades in relative motion. Other flow geometries needed to be analysed as the great liquid-fuelled rocket engines began to be designed. In many such developments the fluid dynamics of *combustion* played a critical part; while, once again, the associated cooling systems needed careful study by fluid-dynamic heat transfer theory. The same ingredients (chemical reactions, heat transfer and fluid dynamics, with the dynamics of mixing playing a dominant role) underlie many current developments in process engineering.

In the meantime the modern kinematic and kinetic analysis of machines and mechanisms has continued to make great progress, which has proved especially vital for the increasingly important technology of robotics. Gyro dynamics, again, has led to remarkable developments in control and navigation through the design of sensitive systems combining gyroscopes and accelerometers.

The modern mechanical engineer, to be sure, synthesizes knowledge from many different disciplines; including (for example) an even wider area of materials science than was touched on earlier in relation to structural engineering, and including also large parts of electronics that are essential for purposes both of instrumentation and control. Theoretical mechanics of solids and fluids continues, however, to form the fundamental basis of the mechanical engineer's art.

## 5. External Fluid Dynamics

The interaction of a solid body with an external fluid through which it moves is the subject of external fluid dynamics. Remarkable developments in external aerodynamics (the case where the fluid is air) were needed, in addition to several structural and power-plant developments that have been referred to earlier, to make possible the twentieth century's extraordinary achievements in aeronautical engineering.

Efficient horizontal flight requires the air pressures and viscous stresses around an aircraft to produce a resultant force with a large vertical component or *lift* to balance the weight, and yet with a low horizontal component or *drag* to be overcome by engine thrust. The viscosity of air is small but in external aerodynamics must not be neglected since classical theories on this assumption predict a zero resultant force on any solid body in steady motion through air. Accordingly, the discoveries by Prandtl and others, that well designed external shapes for aircraft could allow all the effects of viscosity to be confined to thin boundary layers and wakes, permitting drag to be low although (when the wakes contained intense trailing vortices) lift could be high, represented a vital first step forward in aerodynamic design; allowing major improvements in performance over that of the crude shapes of aviation's early years.

Later, after improvements in engines and in structural design as well as in aero-dynamics had raised aircraft speeds near to or above the speed of sound, aircraft began to experience an additional component of drag, associated with generation of an acoustic wave often called the 'sonic boom'. The corresponding feature in external hydrodynamics is observed when a ship's speed increases so that it experiences not only a wake-generating drag but also a *wave-making* drag. Wave theory for both acoustic waves in air and surface waves on water has been fruitfully applied to the study of shapes that will postponed, until substantially increased speeds, any significant rise in drag due to wave-making.

Other important parts of external fluid dynamics are concerned with the stability and control of the motion of aircraft through air or of ships or submarines through water. These parts achieved great success through applying the laws of dynamics to the solid vehicle in each case, taking into account fluid forces as affected by the vehicle's movement and by the action of control surfaces such as rudders, ailerons, etc.

At very high speeds, the integrity of an aircraft's structure may be threatened by frictional generation of heat in the boundary layer and this may once more demand careful heat transfer analyses. In the case of spacecraft re-entering the earth's atmosphere the problems are intensified and the analyses need to be at a very refined level.

Ingenious pressure-measuring and flow-visualisation techniques using copiously instrumented wind tunnels and other experimental facilities have been essential to the successes achieved in external fluid dynamics. Recently, the complete probing of a complicated velocity field in a fluid became possible through the development of Laser Doppler anemometry. Yet simultaneous improvements in computational fluid dynamics, based on major advances in fluid dynamic theory, have taken such a form that the balance of a designer's reliance on the twin supports of experimental data and theoretical analysis has continued to shift slowly but progressively in the direction of theory.

## 6. Planetary Sciences

After the astronomers of the sixteenth and early seventeenth centuries had amassed a remarkable body of accurate planetary observations, and Kepler had established a post-Copernican interpretation of these in terms of three empirical laws satisfied by the orbits of planets (and of their satellites), a first great success of the theoretical mechanics introduced by Newton was his demonstration that those laws must be interpreted in terms of an *inverse-square law* of gravitational attractive force. Much later, simultaneous attraction by more than one body was approximately allowed for, and used by Euler, Lagrange, Laplace and others to explain the principal departures from Kepler's laws which more accurate observations had brought to light.

The perturbation theory developed for this purpose proved invaluable after artificial satellites of earth were launched from 1959 onwards. Careful observation of the slow deviation of their orbits from those predicted by Kepler's laws allowed accurate determinations of the *perturbing forces* on a satellite (that is, forces other than simple attraction to the earth's centre). Important geophysical information was so obtained, including the best available data on departures of earth's shape from the spherical, and on winds in those highest parts of the atmosphere which include the orbits of many satellites.

Study of the lower parts of the earth's atmosphere is, of course, the province of meteorology. The invention of the barometer in the seventeenth century had allowed some empirical progress in the study of the weather, but subsequent advances rested firmly on knowledge of the theoretical mechanics of fluids in a *rotating frame of reference* and on improved understanding of the physics of moist air and of radiation. Admittedly, every big step forward in data acquisition, such as the radiosonde balloon, enlarged the potentialities of meteorology as a science; nevertheless, the data were only useful when interpreted in terms of ideas emanating in part from theoretical mechanics. Some of these ideas proved useful also in the study of planetary atmospheres.

The modern meteorologist is able, however, to go far beyond the simple introduction of ideas from fluid dynamics. The global collection of data from radiosondes and from meteorological satellites now permits *initial conditions* to be determined that allow a forward numerical integration of the partial differential equations for atmospheric motions (including the necessary radiation physics) to be carried out in finite-difference form and used successfully to calculate weather conditions a few days ahead. This rapidly developing field is one of today's particularly important areas of application of theoretical mechanics.

Similar stories can be told in two other branches of the study of *the earth's fluid envelope*: Oceanography, with its highly developed study of ocean tides and currents intimately related to hydrodynamic theory; and ionospheric dynamics, including the dynamics of the interaction of the earth's magnetosphere with the solar wind of charged particles emanating from the sun, Increasingly, the mechanics of the entire fluid envelope needs to be treated as a whole.

Solid-earth geophysics is also highly dependent on mechanics and on its well developed *seismological* theories, allowing information about the interior of the earth to be inferred from data gathered by sensitive seismographs on how waves generated by earthquakes propagate through it. Extensive knowledge of the material properties of the solid outer parts of the earth (the *crust* and *mantle*) were derived by these methods; which in addition, demonstrated the existence of the liquid core. Later, it became recognized that the magnetic field of the earth is generated by 'dynamo' motions within the electrically conducting core through mechanisms that in recent times have been in part clarified by magneto-fluid-dynamics. All of these studies of the earth's interior are found valuable also in investigations of the structure of the moon and of the other planets and their satellites.

# 7. Life Sciences

Beyond the widespread applications of mechanics to different branches of engineering and technology and also to planetary sciences that have been briefly mentioned so far, this note abstains from outlining the interactions of mechanics with other physical sciences; principally, because mechanical ideas have long been fully integrated within disciplines such as physics, physical chemistry and astrophysics (admittedly, with various important relativistic or quantum-theory modifications; and, to be sure, as rather modest parts of the whole in each case). Instead, this description of the applications of mechanics is concluded with an indication of how rapidly the life sciences, after, quite a slow start, have in the last decades been infiltrated by ideas from theoretical mechanics.

This was the period during which, first of all, the very diverse modes of animal locomotion were subjected to detailed analysis from the standpoint of the forces required and of how they are exerted by muscles and other motile organs. That work has comprehensively applied the principles of mechanics of solids to all of the different *gaits* (including creeping, walking, hopping, trotting, cantering, and galloping) used in terrestrial locomotion, Simultaneously, *aquatic locomotion* was being studied from the standpoint of hydrodynamics for a huge range of fishes and other vertebrate swimmers, for mobile crustaceans and molluscs, and for flagellar and ciliate microorganisms. Aerodynamics, again, was used to analyse the many and various modes of forward flight and its control, of takeoff and landing, and of climbing, soaring, diving and hovering used by birds, bats, and insects.

The biomechanics of bone and connective tissue was comprehensively investigated, with applications to the design of prosthetic and orthotic devices. Physiological fluid dynamics was developed, too, for the detailed analysis of *blood flow* in the cardiovascular system, of *air flow* in the respiratory tract, and of fluid motions in various specialised organs such as the ear. There are, furthermore, many other excellent possibilities for future applications of mechanics in the life sciences.

To summarize: The International Union of Theoretical and Applied Mechanics utilises its large International Congresses, along with its relatively smaller specialised Symposia and also its Joint Symposia held in cooperation with other international bodies active in engineering and the sciences, for three main purposes: to promote fundamental, mathematical and computational developments in theoretical mechanics; to advance those new experimental techniques that are needed for applied mechanics to be able to make progress in its diverse fields of application; and above all, to move towards the winning of yet more fruitful results from those powerful interactions between these two fields that have been briefly celebrated in this note.

# Current Solid Mechanics Research

**Viggo Tvergaard**

About thirty years ago James Lighthill wrote an essay on "What is Mechanics?" With that he also included some examples of the applications of mechanics. While his emphasis was on fluid mechanics, his own research area, he also included examples from research activities in solid mechanics.

Solid mechanics is a diverse science with roots in traditional research fields while also being an integral part of many entirely new areas. Much work is going on in classical research areas such as structural mechanics, where also many Ph.D.'s in solid mechanics use their expertise to solve complicated mechanics problems in private companies. This includes off-shore structures, large ships, modern bridges, and very tall buildings. Related to this is the analysis of large shell structures, and there is also a big demand for expertise on contact and friction mechanics, which are important research areas with many unresolved issues. Solid mechanics borders to electronics, medical devices, the energy sector, transportation, materials science, physics, and biology.

In the following a few examples are given of larger hot research areas that currently play an important role in solid mechanics research.

**Materials mechanics**. The classical nonlinear material models describe elastic-plastic deformations of solids or creep at high temperatures. These models are extended in many different directions. One is to account for damage evolution in the material and the effect of this damage on the stress-strain relations for the material. This can be small voids or micro-cracks or the evolution of failure between different parts of a multi-component material, such as a composite. In many of these application areas large deformations occur, so that the models must be developed in the context of general continuum mechanics. Size effects in

V. Tvergaard (*1943)
Danish Technical University, Lyngby, Denmark

13

material response is a very active sub-area here, that has spawned numerous research articles. It turns out that materials often have a characteristic length scale, such that the behavior depends on the size of the solid studied. A number of different mathematical formulations of material models are currently proposed that incorporate effects of the size in different ways.

Fracture mechanics is a subject of central importance in the description of material behavior. This involves singular crack-tip fields, determination of threshold values of crack parameters below which cracks will not grow, and studies of crack growth both in brittle solids and ductile solids or in solids where different kinds of damage develops and interacts with a macroscopic crack. Research in this area often makes use of cohesive zone models for the formulation of fracture criteria. Many solids are subjected to cyclic loading where fatigue fracture is the predominant failure mechanism that attracts much research interest.

Research on the mechanics of materials requires detailed knowledge of basic material behavior and therefore some of this research occurs in close collaboration with materials scientists. Recent interest in smaller and smaller length scales has opened research in nano-mechanics and molecular dynamics, where the research interfaces with materials physics.

**Computational mechanics**. As in many sciences, progress in solid mechanics research is closely tied to the fast development of computers. Problems in continuum mechanics often involve the solution of systems of nonlinear partial differential equations, which is possible only due to computers. Finite difference approximations can be used, but in solid mechanics variational formulations leading to finite element approximations are usually found most efficient. Computational mechanics is not just a separate research field, it is an integral part of most of the research in solid mechanics. To do this on a high level, as is needed, the researcher must have a deep insight in mathematical methods of numerical analysis, in algorithms, and in current developments in computer capabilities. And still the purpose of the solid mechanics researcher is not the development of these advanced methods in themselves, but the use of these methods to obtain new insight in solutions of complex problems in solid mechanics.

**Dynamics**. This important research area has its origin in vibration theory, used to avoid resonance failure in structures. Current research involves multibody and vehicle dynamics, chaotic behavior of dynamical systems, the dynamics and control of morphing structures, and friction induced vibrations. There are also relations to mechatronics such as the use of active or semi-active dampers to control instabilities in rotor dynamics. Methods of particle dynamics are developed to represent certain effects in dynamical systems.

Dynamics also includes wave propagation, which plays an important role in a number of applications, such as impact in vehicle collisions, dynamic crack growth, or fault propagation during earth quakes.

**Instabilities in structures**. This area started already with Euler's analyses of column instabilities. The understanding was much improved in the middle of the 20th century by the development of asymptotic methods for post-bifurcation behavior and associated methods for the evaluation of imperfection sensitivity.

Subsequently, mathematical theories for bifurcation and post-bifurcation in elastic-plastic structures were developed. This has led to a lively area of research in structural instabilities, including subjects such as buckling localization, instabilities in thin surface layers where compressive stresses develop, and instabilities in metal foam.

**Biomechanics**. The use of mechanics to understand biological systems is a research area in rapid evolution. Within solid mechanics one area is bone mechanics. Here researchers need to incorporate typical features of living material where properties depend on and develop due to the applied stress fields.

For soft biological materials the constitutive models applied are strongly related to models developed for other soft materials, such as rubber, polymers or gels. This allows for numerical studies of a disease like aortic aneurysms, where localized arterial expansion can be life-threatening, or red blood cells subject to malaria, where the mechanics of cell deformation is central. The soft biological materials typically undergo large deformations, requiring general continuum mechanics in the formulations.

Also, the mechanical properties of cells attract much interest. Here cross-linked networks of the protein actin are important building blocks for the cytoskeleton. The behavior is somewhat related to polymers, and some researchers build this into continuum models that can be treated by the methods of mechanics. To this area belong also new developments in drug delivery systems.

**Manufacturing**. Metal forming analyses have developed strongly in the last couple of decades from the classical use of simple, highly approximate, rigid-plastic methods to the use of modern computational tools in solid mechanics. It is characteristic in these applications, like rolling, metal cutting or forging, that very large strains develop which must be accounted for in the mechanics formulations to be used for the numerical solutions. There are also cases where the size of the specimens formed is so small that size effects have to be accounted for in the constitutive models.

Additive manufacturing, or 3D printing, is a new process that offers interesting possibilities of near net shape forming in very complex geometries. There is focus on better understanding the material properties obtained in this type of process, and the important effect of unavoidable residual stresses.

**Optimal design**. Optimization of the properties is an important goal for all engineering structures. This can mean minimum weight for a given functionality, maximum strength for a given amount of material, or minimum stress concentration to avoid fatigue. Systematic research in this area has been boosted enormously by the development of optimization algorithms such as linear programming or quadratic programming, which allow for efficient computer solutions with many degrees of freedom. Much of the current research is focused on topology optimization, which uses homogenization theory together with optimization algorithms to find optimal configurations with few initial restrictions. These methods are also used to optimize compliant mechanisms, MicroElectroMechanical Systems (MEMS), and materials with extremal properties.

# Current Research in Fluid Mechanics

Tim J. Pedley

Since Sir James Lighthill wrote the above essay "What is Mechanics?" for the first edition of this book, fluid mechanics, like solid mechanics, has seen remarkably rapid growth and evolution. In part this stems from the needs of industry and government and from the needs of other sciences such as biology, physics and earth sciences (for example). Whatever the application, every new problem requires the development of new fluid mechanics, on account of the nonlinearity and intrinsic complexity of the governing Navier-Stokes equations. However, the principal change over the last thirty years has been the totally unpredicted explosion in the computing power available. Theoreticians have been able to simulate larger and larger fluid systems, with ever-increasing resolution and accuracy, using Computational Fluid Dynamics (CFD) to explore and supplement mathematical models in regions of parameter space for which mathematical analysis is impractical. At the same time the enhancement of computer power has permitted new, non-invasive imaging technology and the ability to visualise complete flow fields (at least in the laboratory) from massive quantities of data. Examples of newly developed methods include Ultra-sound Doppler Velocimetry, Laser Doppler Velocimetry, Particle Image Velocimetry and Magnetic Resonance Imaging. A little more detail on some areas is given below.

**Multiphase Flows**. This topic includes bubbles (of gases in liquids), drops (of liquids in gases), flow of liquids and/or gases in porous media, flow of gases and liquids together in pipes, suspensions of small solid or fluid particles in liquids, flow of granular media, and the mechanics of foams. Surface tension forces are often dominant, and the Marangoni stresses generated by thermal or solutal gradients in surface tension may also be important. The break-up into drops of a cylindrical jet,

T.J. Pedley (*1942)
University of Cambridge, Cambridge, UK

© The Author(s) 2016
P. Eberhard and S. Juhasz (eds.), *IUTAM*,
DOI 10.1007/978-3-319-31063-3_3

17

whose instability has been known since Lord Rayleigh in the 19th century, can now be both visualised and simulated with unprecedented resolution in both time and space. The flow and instability of thin liquid films, as on the inside of pipes (in a chemical plant or the lungs, say) or between rollers in the manufacture of steel sheets, plastic films or paper, has become a recognised subject in its own right. It is both practically important and attractive to applied mathematicians because the governing equations can be reduced in order by use of the lubrication approximation. The flow of granular media, in which the contact forces between the grains dominate the fluid dynamical forces, has also become a lively subject in its own right.

**Non-Newtonian Fluid Dynamics**. In classical fluid dynamics the fluid is in general Newtonian, the deviatoric stress tensor being directly proportional to the strain rate tensor; the constant of proportionality (twice the viscosity) may depend on temperature or solute concentration, but not on strain rate. However, many everyday fluids are non-Newtonian, being shear-thinning, shear-thickening or viscoelastic. This is typically because the fluid is a solution of large molecules (e.g. a polymer) or a suspension of colloid particles, and the scientific concern is twofold: what is the constitutive relation between stress and strain rate that should replace the Newtonian relation, and what are the consequences for the flow patterns and the forces exerted on solid boundaries? As in many fields, understanding the macroscopic properties of the fluid from a mechanistic description of its microstructure (molecules or particles and their interaction) is a problem of fundamental importance and great difficulty. Continuum modelling of granular media is particularly difficult when the grains are irregular, and the gap between experimental data and theoretical or computational prediction remains wide.

**Microhydrodynamics**. When the Reynolds number is small, so that inertia is negligible, the Navier-Stokes equations reduce to the Stokes equations. These equations are linear, but nevertheless the field of microhydrodynamics has continued to be a rich source of both interesting problems and physical understanding. There are perhaps three principal sources of interest: the need for microscopic understanding of concentrated colloidal suspensions, as discussed above; the many microfluidic devices currently being developed under the somewhat misleading title of 'nanotechnology'; and the desire to understand the locomotion and interaction of swimming micro-organisms. The study of internal flows, in cells or vesicles, driven by active molecules or external stresses, is another subfield.

**Biological Fluid Dynamics**. As Lighthill has pointed out, this subject can be subdivided into external fluid dynamics—the interaction between living organisms and their fluid environment—and internal, physiological fluid dynamics. In the former category comes both low-Reynolds-number locomotion of micro-organisms and high-Reynolds-number swimming of fish and cetaceans, flying of birds, bats and insects, filter feeding, deformation of plants by ambient flow, etc. In physiology the interest remains primarily in the circulation of the blood, airflow in the lungs, and the flow of other 'biofluids'. Large computational and experimental programmes have been developed to compute the flow and wall shear stress in the complex 3D geometries of arteries and airways, obtained by imaging of individual

subjects, as aids to diagnosis and treatment, and with a view to optimising prosthetic devices such as artificial heart valve and stents.

Like other fields, biological fluid dynamics has changed its focus over the years,
from being a source of nice problems that lead to new physical understanding of
fluid dynamical processes, to being an essential component in developing new
biological understanding of how plants and animals work. This can only be done in
close association with biological experiments, and these too require totally novel
quantitative experimental techniques, often at the cellular or subcellular level.

**Flow-Structure Interactions (FSI).** The study of blood flow in elastic vessels is
just one area in which dynamic interaction between the fluid flow and solid
deformation is important. Biology yields numerous examples of FSI in which the
structures are soft. The more traditional fields of aeronautical or mechanical engineering continue to yield a rich diversity of FSI problems with stiff structures—
structural stability, flutter, aeroacoustics—that need to be solved in the relevant
industries.

**Geophysical and Environmental Fluid Dynamics (GEFD).** Laboratory
experiments on rotating and stratified fluids have been successfully performed and
understood theoretically. The problem with extending that understanding to the
terrestrial scale—oceans and atmosphere—or larger lies in the difficulty of
achieving enough observational data. Satellite imaging using radiation of various
wavelengths (not just visible light) means that coverage of the atmosphere, especially where it contains particles or water droplets, and of the surface layer of the
ocean, is now extensive, so that the spatial resolution is just about adequate for four
or five day computed weather forecasts. However, data at depth in the ocean are still
sparse, and prediction of ocean behaviour remains an uncertain science.

A major interest in GEFD is in stirring and mixing, from the point of view both
of controlling pollution and of understanding the supply of nutrients to biological
organisms in the ocean. Regions of high solute and particle concentration are
advected with the ambient flow, but are sheared by velocity gradients, so high
concentration gradients develop laterally, permitting diffusive mixing. This, together with the desire to understand turbulence, is one motivation behind the development of vortex dynamics and topological fluid dynamics as a separate subfield of
fluid mechanics.

**Hydrodynamic Instability and Transition to Turbulence** are topics that have
interested fluid dynamicists for over 150 years, and continue to occupy many of us.
Every topic referred to above gives rise to stability problems, which are treated
using traditional linear and weakly nonlinear methods as well as Direct Numerical
Simulation. One exciting development in particular concerns the transition to turbulence in unidirectional (or nearly so) flows in pipes and boundary layers.
Stimulated by remarkably precise experiments on pipe flow, there has been an
extremely fruitful symbiosis between CFD and dynamical systems theory, in which
exact travelling-wave solutions of the Navier-Stokes equations are regarded as
unstable fixed points in the space of all solutions, around which the trajectory of the
actual solution can be understood. Research on turbulence itself continues to

occupy many researchers, but even with modern computer power DNS still cannot simulate turbulence at Reynolds numbers of practical importance!

The above omits all mention of many central and long-standing areas of fluid mechanics, such as the effects of compressibility, or chemical reactions (e.g. in combustion), or heat transfer, or water waves, or magnetohydrodynamics, for example, but it is hoped that the limited survey may be seen as a representative snapshot of current research in fluid mechanics.

# About the Early International Congresses of Applied Mechanics

Giovanni Battimelli

## A. Summary

Several authors have discussed the conflict, during and after the first world war, between the internationalist ideology of scientific knowledge and the political commitment of scientists, in particular with regard to the policy of the International Research Council and of its scientific unions. A case study is presented here of an international body which was born during the Twenties (when the polemic between scientists on opposite sides was at its peak) and quickly attained unpredicted success. Preceded by an informal gathering organized by T. von Karman and T. Levi-Civita in Innsbruck in 1922, the International Congress of Applied Mechanics, first held in Delft in 1924, was, at the end of the decade, much more of a live institution than many of the unions tied to the IRC.

Two factors seem to be especially responsible for this success: On one hand, the programmatic refusal by the "founding fathers" to establish any formal connection between the Congress and any official body or institution tied to the IRC or to single governments, in order to avoid the obstacles of international scientific diplomacy. This "refusal of politics" proved to be an extremely successful political act. On the other hand, the specific nature of the discipline involved has to be taken into account. The International Congress of Applied Mechanics is seen as the body which comes to identify a new sector of scientific activity, bordering on physics, mathematics, and engineering, which took shape in those years mainly in the German-speaking scientific world.

It may be recalled that proposals advanced at the IRC to establish international cooperation on technical matters, tied as they were to old-fashioned disciplinary

G. Battimelli (*1948)
Universita degli Studi di Roma "La Sapienza", Rome, Italy

© The Author(s) 2016
P. Eberhard and S. Juhasz (eds.), *IUTAM*,
DOI 10.1007/978-3-319-31063-3_4

21

subdivisions, and generally inspired by anti-German prejudice, totally failed to materialize.

# B. Prehistory: Innsbruck Conference

In April 1922, the director of the Aerodynamical Institute of the Aachen Polytechnic, Theodore von Karman, wrote to the Italian mathematician Tullio Levi-Civita in Rome, asking for his advice and collaboration on a project he was considering. Karman noted the contrast between the rapid development taking place and the interesting new results being obtained at the time in the field of hydro- and aeromechanics, on one hand, and on the other the little space devoted to such problems at scientific meetings, and the scarcity of personal contacts between scientists engaged in the field. Moreover, Karman added, people interested in fluid mechanics attended either mathematical, or physical, or technical conferences thus limiting even more the possibility for closer interaction. Time was ripe, in Karman's judgement, to break this dependence from the mother disciplines and to give hydro- and aeromechanics the independent status they deserved. How would Levi-Civita assess the prospect of calling an informal meeting of people interested in the field, both from the theoretical and the experimental sides? Karman suggested holding the meeting in Innsbruck in the fall and asked for the Italian's collaboration as co-organizer.

Karman had more than one reason for choosing Levi-Civita, among others, as a partner for the enterprise, apart from their good personal relations. To obtain the best result in terms of international participation, it could be a good move to have as one of the organizers, besides the Hungarian-born von Karman, a leading scientist from one of the Allied Powers, well known for his pacifist and internationalist views. Karman could take care of securing the attendance of German scientists and Levi-Civita could smooth hesitation and resistance that had to be expected from the side of the wartime victors, especially from the French. Also, apart from diplomatic considerations, Levi-Civita was a sound scientific choice. He was basically a pure mathematician; his association with Karman as promoter of a conference on technical matters was a living example of how Karman meant to develop the field, maintaining close connections between technical developments and the empirical work and the more fundamental theoretical research and a high level of mathematical sophistication.

Levi-Civita's answer was warmly enthusiastic, and he especially noted the importance of obtaining as international a participation as possible. "I think"—he wrote back to Karman—"that we should send the invitation also to some English and French scientists... Should any of them actually come, so much the better; in any case it will be clear that we are moved by a spirit of scientific internationalism."

With the material organization of the meeting in his and Levi-Civita's hands, Karman looked around for other renowned scientists to establish a sort of scientific committee and sign the invitation to be sent out. A draft of the invitation letter can be seen in the following. It was easy for him to secure the adhesion of his former teacher, Ludwig Prandtl from Göttingen, but with an eye to diplomacy he felt that he needed someone from the former neutral countries. Writing to Stodola, the turbine expert in Zurich, he said that "political views should be completely bypassed by the fact that the meeting will not represent any official congress, but will be held as a totally informal gathering."

Monsieur et cher collègue,

Un groupe de savants et d'ingénieurs s'occupant d'Hydrodynamique et d'Aeromécanique a pais l'initiativ d'une petite réunion scientifique de personnes intéressées à cette branche de la science.

Dans tous les pays, l'Hydrodynamique a fait dernièrement des progres importants, tant ce qui concerne les théories classiques tant en éluoidant et approfondissant les problèmes fondamentaux de l'Hydraulique pratique et de l'Aerotechnique; mais les événements politiques ont empêché l'échange des idées et des expériences.

Cette petite réunion para sera destinée à ressembler quelques savants et ingénieurs intéressés à ces problèmes afin d'échanger leurs vues scientifiques.

On fera des conférences sur les progrès réalisés dernièrement dans les différents pays et f on les fera suivre de descussions.

Nous avons l'honneur de vous invitez à prendre part à cette réunion qui aura lieu à Irnebruck (Tirol) du 10. au 13. September 1922.

Si vous approuver notre idée, veuillez envoyer un mot à M.Levi-Cività, jusqu'au 20    Août St.Ulrich (Gröden) (Alto Adige, Italie) Hotel Post, après Padova Via Altinate où à M.de Kármán, Institut aerodynamique Aachen (Allemagne). M. de Kármán se tient à votre disposition pour tous les renseignements concernant l'itinéraire, logements etc. Vous receverez bientôt un programme détaillé avec les titres des conférences et les sujets des discussions.

Veuillez aggréer, Monsieur et cher Collègue, l'assurance de nos sentiments très distingués.

T.Levi-Cività    C.W.Oseen    L.Prandtl

Th.de Kármán

In fact, however, Karman worried about possible difficulties arising from political views much more than he was willing to admit. The letter he sent to H.A. Lorentz clearly shows it: "It would be of great importance,"—he wrote to the leading Dutch scientist—"that the invitation to this informal conference, which by no means will have the character of an official congress, be signed by some scientist from the neutral countries... May I stress that, even if by misfortune you could not come to Innsbruck, it would be important for us to have, all the same, your signature." Lorentz was, in fact, prevented from attending the meeting, and therefore politely declined the invitation to sign the conference announcement, which finally was signed by Karman, Levi-Civita, Prandtl, and the Swedish physicist, C.W. Oseen, from Uppsala. Meanwhile, Levi-Civita was trying to obtain the adhesion to the conference of "Mr. Italian Science," Vito Volterra, writing him a letter that presented the whole thing as having been originated by Oseen, with the Germans being there by mere chance. Volterra's strong nationalistic views were well known, and he was at the time the leader of the Italian delegates to the International Research Council, whose policy was primarily inspired by anti-Germanism. His adhesion would obviously have been of great significance, but Volterra did not fall into Levi-Civita's trap and rather harshly replied that he could neither attend the meeting nor give his adhesion.

The difficulties that the two organizers expected began to appear. Two quotations will suffice to give an idea of the atmosphere. On one hand, Marcel Brillouin replied to Levi-Civita: "Meetings such as this, even more than the international congresses, require as a condition the absence of any suspicion that might hinder cordiality. As far as I am concerned, until the German scientists and professors will understand that, in order not to pay for reparations, one should first of all not have caused systematic destruction—and that such destruction having been caused, the Germans must pay for them—my esteem for their moral character is not high enough to allow me to shake hands with them, whatever their indisputable scientific merits." And on the other hand, Richard von Mises wrote back to Karman: "You will not be surprised by the fact that I will not come, since my views about Tyrol and the Italians are well known to you... Furthermore, I am a bit surprised at seeing that German professors feel the need to communicate abroad their theoretical researches on flight, while we are at the same time prevented from building real airplanes."

These statements are quite typical of the sort of accusations and complaints that bounced at the time from each side against the other. They were not, however, the only kind of response with the exception of France and Britain, where clearly people felt that, despite Levi-Civita's presence, the whole thing was too firmly in German, or German-controlled, hands. Elsewhere, the general reaction was a favorable one, and the final list of adhesions included most of the outstanding names in the field. Meteorologists Bjerknes and Ekman came from Norway. Sommerfeld, though prevented from attending in person, sent his student, Heisenberg, to report on the research on turbulence he was doing for his dissertation. Even among the French scientists, there were those who sent messages of sympathy and adhesion, while lamenting that circumstances at home made it impossible for them to be present at the meeting. As can be seen on the displayed handwritten list of participants, thirty-three scientists gathered in September in Innsbruck, coming from Germany, Austria, Holland, Scandinavia and Italy, for what was unanimously regarded as a highly successful conference.

## Hydro-aerodynamische Konferenz
### Innsbruck 1922

| Name | Wohnort | Wohnung in Innsbruck |
|---|---|---|
| V. Walfrid Ekman | Lund | Grauer Bär Zimmer 73 |
| Ludwig Hopf | Aachen | Burgerese |
| Modesto Panetti | Torino (Italien) | Grauer Bär Zimmer 62 |
| F.D. Cycaud | Helmond Holland | " " " 82 |
| Ludwig Schiller e.i. | Leipzig | Tyrol |
| C. Wieselsberger | Göttingen | " |
| G. Reuss | Hamburg | Sonne. |
| W. Kisenberg | München | " |
| K. Ludloff | München | Andreas Hofer. |
| T. Levi-Civita | Roma | Karlberger Hof (№62) |
| J. Th. Thysse | Haag | Grauer Bär Zimmer 101 |
| B.G. v.d. ... | Amsterdam | Grauer Bär Zimmer 104 |
| J.M. Burgers | Delft | 92 |
| Pronto Caldonazzo | Milano | Grauer Bär 66 |
| Th. Josephine v. Kármán | Aachen | " " 44 |
| M. Eisotti | Milano | " " 65 |
| Philipp Frank | Prag | Grauer Bär 79 |
| V. Bjerknes | Bergen | " 146 |
| H. Föttinger | Danzig | " " 126 |
| ... | ... | ... 3 |
| W. ... | ... | ... |
| Tagger | Innsbruck | Hötting 18 |
| L. Prandtl | Göttingen | Grauer Bär 112. |
| Ph. Forchheimer | Wien | Grauer Bär 7 |
| T. Pöschl | Prag | " 2 |
| G. Leskowiz | München | " 139 |
| D. Thoma | München | K. 1. |
| C.W. Oseen | Uppsala | " № 64. |
| C. Tumlirz | Innsbruck | Speckbacherstr. 26 |
| E. Treffte | ... | Bürmese |
| C. ... | Amsterdam | Grauer Bär 76 |
| J.M. Baumham | ... | " |
| J. De Marchi | Rom | ... Hof |

## C. First Congress—Delft

Karman looked forward. Innsbruck was to him but the first step toward a more ambitious goal. He found a new, enthusiastic partner in one of the young scientists who had been at Innsbruck, the former student of Ehrenfest, Jan Burgers. They discussed the prospect of calling a wider conference, no longer restricted to hydro- and aerodynamics, but embracing the whole field of applied mechanics. The Technical University of Delft, where Burgers held the chair of fluid mechanics, was identified as a proper place for the meeting, located, as it was, in a neutral country. Burgers and his friend and colleague Biezeno, the expert in elasticity theory, began again, with Karman's collaboration, the diplomatic ballet to form an Executive Committee as international as possible. Again, the winning card was expected to be the reasserted totally informal character of the conference. This worked; in fact, the number of nations represented in the Executive Committee grew from the six who were present in Innsbruck to ten, particularly notable being the adhesion of Ames and Hunsaker from the United States and of Taylor, Stanton, Griffith, and Coker from Britain.

Problems arose, however, with the two main antagonists, Germans and French. Invitations were sent to join the Committee to four French scientists, who either ignored or refused the proposal. Meanwhile, Prandtl and von Mises let it be known that they did not want any contact with the French, and they stated this as a condition for their adhesion to the Committee. Karman was rather upset by what he termed the "Katz- und Mausspiel" (cat and mouse game) between German and French scientists and tried quite firmly to let Prandtl understand that not only he did not share his teacher's and von Mises' views, but that he also regarded their position as being politically and scientifically shortsighted and wrong. "I wish to remark"—he wrote to Prandtl in December 1923, "that if the Congress will be held under these circumstances (one could say with the participation of all nations but France), this would represent an essential improvement for the official recognition of German science in the whole world, and I think that to withdraw in this case would be a completely wrong move from our side."

The solution of the impasse came as a result of the spontaneous withdrawal of the French. However, despite their absence, the Congress held in April 1924, turned out to be an unexpected success. It was made more remarkable by the fact that even a Belgian scientist attended the Congress (and was immediately co-opted into the committee). Over 200 scientists were present at Delft, where 76 papers were read, spanning the whole spectrum from mathematical problems in rational mechanics to experimental results in material strength. Some of these papers still stand as landmarks in the history of applied mechanics, such as the work by the Russians Friedmann and Keller on the closure problem for the hierarchy of the hydro dynamical equations. Friedmann, too, was invited to join the committee, which established itself as the International Congress Committee.

# D. Second and Third Congresses

It was decided that the Congress would assemble every fourth year, with the second to be held in 1926 in order to obtain a differential of two years with the International Congresses of Mathematics. Again, stress was laid upon the fact that the Committee members acted only as individuals and by no means represented any institution; no official designations by academies or governmental agencies would be accepted, and the Committee would only grow by cooptation. And soon it grew: by the end of 1925, following a renewed invitation, four French scientists joined the Committee. (It may be recalled that only in June 1926, a resolution was voted by the IRC General Assembly to invite the former Central Powers to join an invitation that, by the way, was refused by the Germans).

Meanwhile, difficulties started to appear regarding the choice of the site of the next Congress. Both Italy and Switzerland had offered to host the meeting in 1926, in Rome and Zurich, respectively. Pressures exerted by the Germans, mainly by von Mises, led to a preference for Zurich as a safer choice, Switzerland being a former neutral country. Levi-Civita, therefore, renewed the invitation to host the Congress in Rome in 1930, but the deterioration of political relations between Italy and Germany at the time again led von Mises to remark that a decision to hold the meeting in Rome could possibly result in a boycott by the German scientists. He asked Karman to exert pressure on Levi-Civita to convince him to withdraw the invitation. Karman refused to do so; if there was anyone he didn't want to irritate, he replied, it was the Italians because Italy was the only former enemy country that had been in on the venture from the very beginning. Finally, an invitation was offered by Oseen to hold the meeting in Stockholm, and Sweden was chosen, with general relief, as the country that would host the 1930 Congress.

To raise spirits again, the Belgians came forward with a new idea: in 1930, an International Exposition would be held in Liege to celebrate the hundredth anniversary of Belgium's independence, and it was suggested, too, that scientific meetings be held on the same occasion. The Rector of the University of Liege, therefore, asked Oseen to turn over to the Belgians the organization of the third Congress. Reaction was strong this time, not only, as was to be expected, by the Germans, but by virtually everyone in the Committee. This proposal seemed, in fact, to endanger one of the basic rules the founding fathers had decided to adhere to, namely, the totally non-official character of the Congress. Commenting on this point in a letter to Oseen, Burgers and Biezeno remarked, "In total contrast with the Belgians, who seem to regard a political event as a good reason to call for a scientific meeting, we think that no occasion would be less appropriate to this purpose, if the aim is really to promote the personal cooperation between scientists of different countries... We cannot see in this proposal anything else but a destructive action against the principles and the goals of our organization."

The 1930 Congress, which eventually gathered in Stockholm, marked a peak of attendance, with about 600 scientists. At the end of the meeting, sixteen nations, virtually the totality of the scientific world, were represented in the International

Committee. Over a decade characterized by exclusion and conflicts between scientists and scientific institutions on opposite sides, the original project of von
Karman, Levi-Civita, Burgers, and their colleagues had developed into an
astoundingly successful international enterprise.


# E. International Relations in Sciences After WWI

The story of this particular success seems to lend further evidence in support of
some general features of international relations in science after World War I that
have been pointed out by several authors in recent times. These include the fact that
informal gatherings and freelance organizations had a chance of success, whereas
the resumption of official international collaboration was blocked by political and
diplomatic obstacles. Also the clash between the alleged international character of
science and the national loyalty and political views of individual scientists could
more easily be resolved when national and political considerations were, formally at
least, left outside the door. Finally, the scientists from the neutral countries, mostly
those connected with the German-speaking scientific community, played a fundamental role in leading Germany back into the international network.


# F. Nature of the Scientific Discipline: Applied Mechanics

However, although the wise political conduct of the enterprise by Karman and
colleagues may seem to suffice to account for the success met by their ambitious
plan, a few more remarks about the nature of the specific discipline involved are in
order. We are dealing here with applied mechanics, a discipline that simply did not
exist, as a field in itself, before World War I. It was not simply a matter of resuming
the international collaboration disrupted by the war in some established scientific
field. In this case a new sector of the international scientific community was shaping
its own identity by the very foundation of the Congresses. It is not surprising that the
original idea of such a congress and the driving forces behind it came from scientists
either in Germany or in those countries more closely connected to German scientific
circles; it is in Germany that the need for, and the first steps toward, the establishment of applied mechanics as a separate field in itself began to appear right at the end
of World War I. During and after the 1920 Naturforscherversammlung in Nauheim,
some of the leading exponents of the field (Prandtl, von Karman, von Mises, Trefftz)
had exchanged ideas and agreed on the conclusion that separate sessions for applied
mechanics were needed at these meetings, distinct from those for mathematics and
physics. This first happened the following year at the Naturforschertagung in Jena,
which was prominently reported in a new journal that had appeared a few months

earlier—the Zeitschrift für Angewandte Mathematik und Mechanik, edited by von Mises in Berlin, a journal that was soon to become the main reference and a favorite place of publication for researchers in that field.

There is abundant evidence, in the correspondence and in the personal recollections of the scientists we are discussing here, that they distinctly regarded themselves as the exponents of a new science, placed somewhere at the intersection between mathematics, physics, and engineering—the established disciplines in which most of them had their former training as scientists. The establishment of an international organization for applied mechanics also meant to emphasize this newly acquired identity of the discipline, going beyond the borders of Central Europe and giving it a marked worldwide character. Throughout the early life of the International Congresses, the founding fathers watched carefully that the scientific organization of the meetings should reflect the fundamental characteristic they meant to impress on their discipline, which may be summed up in the close connection between applied goals and theoretical research, empirical work and mathematical investigation: "turning engineering design into engineering science," as Karman used to put it. The 1924 meeting in Delft was divided into three main sections, one on hydro- and aerodynamics, the second on theory of elasticity, and the third on rational mechanics—thus stressing the importance of more theoretically oriented mathematical research for the development of the sectors more closely tied to practical applications. Two years later, when it appeared that the local Committee in Zurich meant to alter the name from "Applied Mechanics" to "Technical Mechanics" Burgers and Biezeno hastened to let Meissner know how much they disliked this prospect: "We have actually discussed the thing in detail in 1923, and we believe that the name now in use is to be preferred, because it embraces a wider field. For the development of mechanics, it is very important that the connection with the contiguous fields of mathematics and physics be maintained."

# G. Epilogue

The advocacy of a tight link between the different facets of mechanics (always looking for the applicative goals, never neglecting the theoretical foundations) was clearly reflected throughout the inter-war years in the organization of the Congresses. A quick glance at the list of papers presented at each Congress will suffice to show how fundamental new theoretical results were reported alongside with significant technological advances. This spirit continued to be manifest in later years. In the mid-forties, the scientists had to face again a variety of obstacles to revive international collaboration in the new climate following the end of the second world war, with a changed political, institutional, and scientific environment; however, the basis on which the new scientific institution was formed remained the same. When the International Union came into existence, the adding of "Theoretical" to "Applied Mechanics" aptly stressed the fact that the old mark that had been impressed from the very beginning was still there.

# Post World War II Activities: From Paris to Lyngby

Jan Hult and Nicholas J. Hoff

Science knows no national boundaries. International contacts have always existed between scholars and scientists in various countries. The First World War interrupted much of this fabric. It took time for bitterness and distrust to disperse. Gradually the wounds were healing. The successful International Congresses of Applied Mechanics, arranged at four year intervals, had established a tradition to be followed in years ahead. At the 5th Congress, held in Cambridge, Massachusetts in 1938, it was decided to hold the next congress in Paris in 1942.

## A. The Union is Formed

The disruption of international scientific cooperation caused by the Second World War was deeper than that caused by the first war. The need for reknotting ties seemed stronger than ever before when the mechanics community reassembled in Paris for the Sixth Congress in 1946. Several countries were absent, as was true of the congress two years later in London. By 1956, practically all the traditional participating nations were united again in Brussels.

As in various other branches of science, international unions had been formed for promoting cooperation in astronomy, chemistry, crystallography, geodesy and geophysics, geography, physics, scientific biology, and scientific radio. The International Council of Scientific Unions (ICSU) was created to coordinate various activities

J. Hult (1927–2013)
Chalmers University of Technology, Gothenburg, Sweden

N.J. Hoff (1906–1997)
Stanford University, Stanford, USA

© The Author(s) 2016
P. Eberhard and S. Juhasz (eds.), *IUTAM*,
DOI 10.1007/978-3-319-31063-3_5

among the unions and to form a tie between them and the United Nations Educational, Scientific, and Cultural Organization (UNESCO). Paris was the site of ICSU as well as of UNESCO.

Under these circumstances, it seemed an obvious step, at the Sixth Congress of Applied Mechanics in Paris, to strengthen bonds by forming an international union on the same patterns as those already existing. Hence, IUTAM was created, statutes were adopted, and the union was admitted to ICSU in 1947.

Whereas the International Committee for the Congresses of Applied Mechanics (ICCAM) consisted of individuals representing only themselves, the union was formed by organizations active in scientific work in theoretical and applied mechanics. The ICCAM itself was one such organization; others were national bodies representing scientists in theoretical and applied mechanics.

The nucleus of the Union Council later named its General Assembly, naturally consisted of ICCAM members because very few national organizations were in existence at the formation of IUTAM. The executive body (Bureau), elected in 1948, consisted of:

J. Pérès, France, President
R.Y. Southwell, England, Vice President
J.M. Burgers, Holland, Secretary
H.L. Dryden, USA, Treasurer
F.H. van den Dungen, Belgium, Member
J. Nielsen, Denmark, Member
H. Favre, Switzerland, Member
G. Colonetti, Italy, Member

One of its first duties was to encourage colleagues in various countries to form national organizations that might join IUTAM as Adhering Organizations. In the first five years of the union, the following national organizations were admitted:

1948   The Royal Society of London
       The Hungarian Academy of Sciences, Budapest
1949   Le Comité National Francais de Mécanique, Paris
       The National Committee on Theoretical and Applied Mechanics of the Czechoslovak National Council of Researches, Prague
       The National Committee on Theoretical and Applied Mechanics of the Norwegian Academy of Science and Letters, Oslo
       Consiglio Nazionale delle Ricerche, Rome
       Le Comité National de Mécanique Theorique et Appliquée de la Classe des Sciences de l'Académie Royale de Belgique, Brussels
       The U.S. National Committee on Theoretical and Applied Mechanics, New York
       The Academy of Technical Sciences of Denmark, Copenhagen

1950   The Swedish National Committee for Mechanics, Stockholm
        The Turkish Society for Pure and Applied Mathematics
        The Ministry of National Resources and Scientific Research, New Delhi
        L'Ecole Polytechnique Fédérale, Zurich
        Die Gesellschaft für angewandte Mathematik und Mechanik, Stuttgart
        Israel Society for Theoretical and Applied Mechanics, Haifa
        Instituto Nacional de Tecnica Aeronautica, Madrid
1951   Die Österreichische Akademie der Wissenschaften, Vienna
        The National Committee for Theoretical and Applied Mechanics of the
        Science Council of Japan, Tokyo
1952   Le Comité National de liaison avec l'IUTAM du Conseil des Academies de
        la RFP de Yougoslavie, Belgrade
        De Nederlandes Commissie voor Theoretische en Toegepaste Mechanica,
        Delft
        The Finnish National Committee on Mechanics, Helsinki
        Polskiej Akademii Nauk, Warsaw

In 1984, the number of National Adhering Organizations was thirty-six.

The General Assembly was to be composed of representatives of the National Adhering Organizations and also of personal members elected by the General Assembly itself. This somewhat unsymmetrical arrangement caused much discussion in the first years, and during its meeting in Pallanza, Italy in 1950, the General Assembly made an amendment to the statutes to the effect that only one Adhering Organization be recognized for each country, and that personal membership be reserved for exceptional cases only.

A decision taken at Paris in 1946 to hold the next Congress four years later was changed when it became known that the next International Congress of Mathematicians would be held in 1950. An invitation from Great Britain to host the Seventh Congress in London in 1948 was then accepted. Since then, the Congresses have been held in Istanbul (1952); Brussels (1956); Stresa, Italy (1960); Munich (1964); Stanford (1968); Moscow (1972); Delft (1976); Toronto (1980); and Lyngby, Denmark (1984).

Since 1972, the title has been International Congress of Theoretical and Applied Mechanics, to conform with the title of the Union.

From the creation of IUTAM, the International Committee for the Congresses of Applied Mechanics, which was formed during the First Congress in Delft in 1924, had existed as an independent body. It was an autonomous member of the IUTAM General Assembly with voting rights. This arrangement was terminated in 1964 when the International Committee was dissolved. Instead, IUTAM established a standing Congress Committee within the Union, which was given responsibility for arranging future Congresses. Members of the Congress Committee were appointed by the General Assembly.

# B. IUTAM Symposia

One of the reasons for forming the Union in 1946 was the desire to increase cooperation in mechanics research. The financial means of the International Committee for the Congresses of Applied Mechanics had been extremely limited, and no activities outside the quadrennial Congresses were possible. With regular annual payment of national membership dues, and with a subvention from UNESCO, there followed the possibility to arrange various specialist meetings between the Congresses. Such "colloquia", later named IUTAM Symposia, have subsequently become a dominating part of the union's activities. The first such meeting, on "Problems of Cosmical Aerodynamics," was arranged jointly with the International Astronomical Union in Paris in 1949. Such cooperation with other unions within ICSU has proved to be beneficial on many later occasions.

The number of IUTAM Symposia held every year has gradually increased from two or three in the first twenty years to about eight in the 1980s.

In an early period of the Union, subjects for symposia originated within the Bureau, but as the symposia activities expanded, the General Assembly began to exert increasing influence on the process. The number of proposals for IUTAM Symposia increased steadily and, in 1977, two panels were set up to scan proposals made by members of the General Assembly in the fields of fluid and solid mechanics. Through the work of these two panels, the General Assembly has an efficient means of creating suitable symposium programs.

In contrast to the International Congresses of Theoretical and Applied Mechanics, the IUTAM Symposia have been reserved for invited scientists only. This has made it possible to limit the number of participants, ensuring efficient work and lively discussion. The scientific committee appointed for each IUTAM Symposium has the duty to edit and publish proceedings from the symposia, making the results known to all experts in the field. In spite of this, certain criticism has occasionally been aired at the closed shop system from scientists not invited.

# C. The Circuit Widens

In 1969, two international organizations (the "International Centre for Mechanical Sciences" and the "International Centre for Heat and Mass Transfer") approached IUTAM proposing that they be affiliated with the Union.

Because the statutes of IUTAM at that time neither contained nor excluded the possibility of affiliation of other scientific organizations, an amendment of the statutes was required. The Bureau received the inquiries with great interest and decided to propose changes in the statutes to the General Assembly that would define conditions for such affiliation.

In 1970, the General Assembly amended the statutes by an article defining the conditions for other international organizations engaged in scientific work closely related to that of the Union to be affiliated with IUTAM. Since then, the following organizations have been affiliated:

International Centre for Mechanical Sciences (CISM), 1970
International Centre for Heat and Mass Transfer (ICHMT), 1972
International Committee on Rheology (ICR), 1974
European Mechanics Committee (Euromech), 1978
International Association for Vehicle System Dynamics (IAVSD), 1978
International Society for the Interaction of Mechanics and Mathematics (ISIMM), 1978
International Congress on Fracture (ICF), 1978
International Congress on Mechanical Behaviour of Materials (ICM), 1982
Asian Fluid Mechanics Committee (AFMC), 1982
International Association for Computational Mechanics (IACM), 1984

# D. Closure

From its birth in Paris in 1946 to the 16th International Congress in Lyngby in 1984, IUTAM has shown a steady growth and is now of worldwide extent. Despite this expansion, IUTAM has managed to be a forum for fruitful person-to-person contact and scientific exchange, largely due to the careful observation of article IV of the Statutes according to which the General Assembly shall be guided by the tradition of free international scientific cooperation, developed in the International Congresses for Theoretical and Applied Mechanics. It is hoped that this spirit of cooperation will continue in the future, and IUTAM will operate solely for the advancement of mechanics and the good of mankind.

# Turn of the Century Activities: From Grenoble to Montreal

Werner Schiehlen

A first report on the activities of IUTAM, the International Union of Theoretical and Applied Mechanics, covering the first forty years of the Union from 1946 to 1986 was composed by J. Hult and N.J. Hoff, and incorporated in the first edition of this book edited by S. Juhasz. This update of IUTAM's history edited by P. Eberhard is devoted to the next thirty years from 1988 to 2016 including the turn of the century. In the following, the developments are described which came up during that period.

## A. Adhering Organizations

Organizations of scientists in theoretical or applied mechanics that effectively represent independent scientific activity in a country or in a definite territory can be admitted by the General Assembly as Adhering Organization of the Union. Each adhering organization has representatives in the General Assembly of the Union with voting rights, and pays annual dues to the Union.

The number of Adhering Organizations admitted by IUTAM during the last thirty years increased from 37 to 48 in the year 2015. The number has been changing due to new admittances or suspensions, respectively, as follows.

W. Schiehlen (*1938)
University of Stuttgart, Stuttgart, Germany

© The Author(s) 2016
P. Eberhard and S. Juhasz (eds.), *IUTAM*,
DOI 10.1007/978-3-319-31063-3_6

37

## Admittances

- 1988 Saudi Arabia
- 1990 Vietnam, Korea
- 1994 South Africa
- 1996 Chile, China—Hong Kong
- 2008 Mexico
- 2010 Cyprus

## Changes of Territories and Suspension

- 1990 Unification of Germany, then FRG and GDR are jointly represented again
- 1992 Dissolution of USSR, then separately represented by Russia (1992), Estonia (1992), Georgia (2000), Ukraine (1995)
- 1993 Dissolution of Czechaslowakia, then partially represented by Czech Republic (1993)
- 1994 Dissolution of Yugoslavia, then separately represented by Croatia (1994), Serbia (2003), Slovenia (1994)
- 2013 Suspension of Argentina

For more details see the appendices. The current list of Adhering Organizations is also available at IUTAM's website.

# B. Affiliated Organizations

International organizations mainly occupied in fields closely related to that of IUTAM can be admitted by the General Assembly as Affiliated Organization of the Union. Each Affiliated Organization has the right to appoint an observer, who is invited to take part in the General Assembly without voting rights. The IUTAM Bureau has the reciprocal right to appoint a nonvoting observer to the corresponding council or other executive body of the Affiliated Organization.

The number of Affiliated Organizations admitted by IUTAM during the last thirty years increased from 10 in 1984 to 20 in 2015. The following organizations have been newly affiliated in this period.

- Latin American and Caribbean Conference on Theoretical and Applied Mechanics (LACCOTAM), 1992/2010
- International Association for Boundary Element Methods (IABEM), 1994
- International Society for Structural and Multidisciplinary Optimization (ISSMO), 1996

- International Association for Hydromagnetic Phenomena and Applications (HYDROMAG), 1996
- International Institute of Acoustics and Vibration (IIAV), 1997
- International Commission for Acoustics (ICA), 1998
- International Congresses on Thermal Stresses (ICTS), 2002
- Beijing International Center for Theoretical and Applied Mechanics (BICTAM), 2010
- International Association for Multibody System Dynamics (IMSD), 2014
- International Association for Structural Control and Monitoring (IASCM), 2014

This list shows clearly the wide extension of fields of mechanics closely related to theoretical and applied mechanics. Thus, IUTAM represents not only most nations considering mechanics as a science but also many related fields beyond fluid mechanics, solid mechanics and dynamics. A current list of Affiliated Organizations is available at IUTAM's website, too.

# C. Changing Activities

The main activities of IUTAM include the organization of International Congresses of Theoretical and Applied Mechanics (ICTAM) through a standing Congress Committee, and the support of IUTAM Symposia and Summer Schools for subjects falling within the field of theoretical and applied mechanics. Moreover, to strengthen the scientific cooperation between IUTAM and its Affiliated Organizations, Working Parties were established for a limited period from 1996 to 2014.

## 1. World Congresses

The scientific program of the congresses followed, until the ICTAM 1980 in Toronto, more or less the same pattern, namely Opening and Closing Lecture at plenary sessions, and invited Sectional Lectures and Contributed Papers in parallel lecture sessions. Major changes were implemented at ICTAM 1984 in Lyngby. To highlight recent topics of special emphasis three Minisymposia within ICTAM were selected. Due to the great success, since ICTAM 1996 in Kyoto six Minisymposia are included in the Congress.

Minor changes resulted in strengthening the invited sectional lectures which are now scheduled exclusively in parallel beginning with ICTAM 2000 in Chicago. The number of these keynote lectures amounted between 12 and 20 during the last thirty years. More details on the World Congresses are listed in the appendices.

## 2. Scientific Symposia

The aim of an IUTAM symposium is to assemble a group of active scientists within a well-defined field for the development of science within that field. In order to achieve an effective communication within this group it is necessary to limit the number of active participants. To this end, all IUTAM Symposia are reserved for invited participants. Those wishing to participate in an IUTAM Symposium are therefore advised to contact the Chairman of the Scientific Committee in due time in advance of the meeting.

Following proposals made by individual members of the IUTAM General Assembly and recommendations given by the two IUTAM Symposia Panels and by the IUTAM Bureau, the General Assembly decides upon topics to be treated at forthcoming symposia. The IUTAM Bureau then appoints a Scientific Committee, which will become responsible for the symposium. The Scientific Committee proposes the date for the symposium, subject to approval by the Bureau.

The Secretary-General of IUTAM informs the Chairman of the Scientific Committee about the Rules and the Guidelines, issued by the IUTAM General Assembly and keeps him continually informed about any decisions regarding his Symposium, which may be taken by the General Assembly or the Bureau. The main responsibility of the Scientific Committee is the selection of scientists to be invited. It is responsible for the publication of the proceedings of the Symposium and for the distribution of the grants by IUTAM to participants.

Due to the great success, scope and format of IUTAM Symposia have not been changed. But their topics widened. In the first forty years the symposia dealt mainly with Fluid and Solid Mechanics, Dynamics and Control as well as Foundations of Mechanics. More recently topics like Bio and Earth Sciences, Fluid Structure Interaction, Materials, Micro- and Nano-Mechanics, Computational Mechanics, and Optimization evolved. Nevertheless, the number of IUTAM Symposia remained due to the strong review process between 7 and 10 per year.

Detailed information on the Symposia is collected in the appendices.

## 3. Summer Schools

The IUTAM Summer Schools were established 1990 in cooperation with the International of Centre for Mechanical Sciences (CISM), our first Affiliated Organization. By tradition, IUTAM sponsors one summer school per year, usually in cooperation with CISM. Exemptions have been Summer Schools in Aalborg, Denmark (1994), Stockholm, Sweden (1995), Beijing, China (2001, 2002 and 2004), Evanston, USA (2010) and West Lafayette, USA (2012).

## 4. Working Parties

At the 1996 General Assembly six IUTAM Working Parties (WPs) were temporarily appointed to initiate a closer cooperation with the organizations affiliated with IUTAM as follows.

- WP 1: Mechanics of Non-Newtonian Fluids (ICR)
- WP 2: Dynamical Systems (IAVSD)
- WP 3: Fracture Mechanics and Damage (ICF)
- WP 4: Mechanics of Material (ICM)
- WP 5: Electromagnetic Processing (HYDROMAG)
- WP 6: Computational Mechanics (IACM)

Then, later the Working Parties were reviewed, one of them dissolved and four new ones added representing timely scientific fields. The 2002 General Assembly approved the following nine Working Parties.

- WP 1: Mechanics of Non-Newtonian Fluids
- WP 2: Dynamical Systems
- WP 3: Mechanics of Material
- WP 4: Material Processing
- WP 5: Computational Mechanics
- WP 6: Biomechanics
- WP 7: Nano- and Macro-Scale Phenomena in Mechanics
- WP 8: Geophysical and Environmental Mechanics
- WP 9: Education in Mechanics and Capacity Building

In 2014, the General Assembly terminated the activities of the Working Parties.

To strengthen the subfields of mechanics, certain working parties may be revived again by commissions which are established with many ICSU unions, too. These kinds of activities may become interesting again for IUTAM in the future.

# D. Statutes, Rules, and Procedures

The original version of the IUTAM Statutes was adopted in 1947 and ICSU, the International Council for Science, accepted IUTAM in the same year as adhering body. The International Council for Science (ICSU) is a non-governmental organisation with a global membership of national scientific bodies (as of 2015 there are 121 members, representing 141 countries) and International Scientific Unions (32 members).

Now, the changes of the Statutes and the Rules of Procedure during the last three decades will be reviewed.

## Statutes

- **1990** In Article VI the Bureau members were added as ex officio members of the General Assembly of the Union. Their terms of membership shall coincide with their term of service on the Bureau.

  In Article XI it was stated that the other members of the Bureau not serving as officers shall have been members of the General Assembly at some time within the four years preceding the time of election of the Bureau.

- **1994** A new Article XV was added stating that an alteration of the Statutes requires proposals either prepared by the Bureau or supported by statements to the General Assembly signed by at least ten of its voting members.

- **2004** In Article II the principle objectives of the Union were extended to scientific work in all branches of theoretical and applied mechanics and related sciences, including analytical, computational and experimental investigations.

  Article VI was extended to distinguish between voting members and observers without voting rights. The voting members are representatives of the adhering organizations, members of the Bureau, and members-at-large. The term of a member-at-large shall be determined by the General Assembly at the time of the election. The following categories of observers are invited to take part in the General Assembly: representatives of affiliated organizations; Secretary of the Congress Committee; chairs of the Symposia Panels; chair persons of the Working Parties; representatives of countries applying for membership; representatives of committees and groups of scientists, if so decided by the General Assembly.

- **2008** In Article VI the representatives of Adhering Associate Organizations have been added to the observer categories.

  A new Article IX was added introducing the adhering associate organisations for countries or territories of the developing world. As a consequence the Articles IX to XIV from 2004 were renumbered as Articles X to XV.

  Article XIV was extended by adding the subscriptions of the adhering associate organizations.

  A new Article XVI was added to specify the subscription payments of the adhering associate organisations.

  As a consequence Article XV from 2004 was renumbered as Article XVII.

- **2014** Article VI was changed. The Secretary of the Congress Committee and the Chairs of the Symposia Panels appointed by the Bureau were moved from the observer category to full members of the General Assembly with voting rights.

  Article XII was specified with respect to the candidates for all seven positions to be elected for the Bureau. These candidates must have been full voting members of the General Assembly.

  In Article XIIIb it is stated that the Secretary of the Congress Committee shall be nominated by the Congress Committee and elected by the General Assembly for a four year term with the possibility of re-election for a second term.

Article XIIIc defines now the election of the members of the Congress Committee from nominations invited by the Secretary of the Congress Committee.

Article XIIId specifies now the election of the Executive Committee of the Congress Committee by the General Assembly.

# Procedures

- **1992** The procedure for electing Members-at-Large of the General Assembly was established.
- **1994** The procedure for election of the Bureau of IUTAM was specified in detail. The Electoral Committee (EC) shall invite suggestions, take account of, and submit to the Secretary General nominations for the candidates for the election to the Bureau: one name for each of the Officer positions (P, S, T) and one or more names for each of the non-Officer positions (W, Y, X, Z). The EC will make sure that the candidates thus nominated are willing to accept an election. The General Assembly shall vote separately on each of the seven positions. In any case in which there is more than one candidate for a position, the vote shall be by secret ballot.
- **2004** The procedure for election of the Bureau of IUTAM was further specified. The invitation of suggestions was restricted to the voting members of the General Assembly, the representatives of affiliated organisations and the Secretary of the Congress Committee.
- **2008** Rules of procedure for Adhering Associate Organisations were established with respect to the prime criteria for the eligibility of any country, and the rights and privileges of Associate Members included.
- **2014** Rules of procedure for the Congress Committee of IUTAM were specified as follows: During an International Congress, the CC shall review proposals for the next International Congress and select the location by a vote of the CC members present (i.e., proxy votes are not permitted). This selection process will typically be accomplished over two separate meetings of the CC.

All these changes of the Statutes and the Procedures are evidence of the IUTAM's leading role in the sciences based on mechanics with many analytical, computational and experimental fields within ICSU, the International Council of Science. The organizational structure of IUTAM has proven to be capable and efficient. Close scientific cooperation and personal long-lasting friendship are reasons for IUTAM's success. The organizational structure of IUTAM is shown in the following figure.

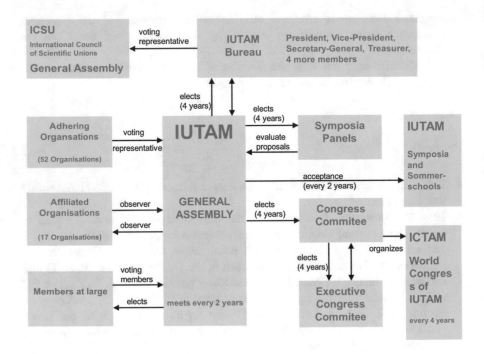

# E. Publications

The first publication considered as an IUTAM publication dates back to the Proceedings of the First International Congress of Applied Mechanics held 1924 in Delft, The Netherlands. It was simply entitled as "Applied Mechanics". Later on in 1972, since the 13th Congress, the Proceedings were entitled as "Theoretical and Applied Mechanics" and the Congress abbreviation ICTAM was used. The congress proceedings were published in cooperation with commercial or academic publishers. In this century, the titles of the Proceedings were more individually chosen as "Mechanics for a new Millenium", "Mechanics for the 21st Century", "Mechanics Down Under" and "Mechanics for the World".

Since the foundation of IUTAM all its activities are documented in Annual Reports in great detail published by the IUTAM Secretariat. The first volume of IUTAM Symposia Proceedings was published in 1949. Symposia proceedings may appear as books or special issues of journals in cooperation with commercial companies. From January 1996 until December 2010, Springer Science and Business Media (named Kluwer Academic Publishers until 2004) has been the preferred publisher of the refereed proceedings of IUTAM symposia under the title IUTAM Bookseries. Since January 2011, Procedia IUTAM by Elsevier is the preferred medium of publication for the refereed proceedings of IUTAM symposia and ICTAMs. It is open access.

An IUTAM Newsletter was established in 1992 and it is distributed widely by the IUTAM Secretariat. The Newsletter contains usually a President's Address, a short report on General Assembly and Congress Committee meetings as well as announcements on the forthcoming ICTAM, IUTAM Symposia and Summer Schools, and IUTAM co-sponsored events.

The first publication on the short history of IUTAM dates back to 1988, it was edited by S. Juhasz and published by Springer-Verlag. On the occasion of the millennium the booklet "Mechanics at the Turn of the Century" was edited by W. Schiehlen and L. van Wijngaarden and published by Shaker Verlag. This report was the result of an initiative of the Bureau of IUTAM to provide some landmarks on the developments in mechanics during the 20th Century, to report on the 50 years of impulse to mechanics by IUTAM, to visualize the history of mechanics by the water colour Meters of Motion painted on the occasion of the 20th ICTAM, to look ahead on a very personal basis, and to show the broad international involvement of scientists in IUTAM in recent years.

The bibliographical details of the broad spectrum of IUTAM related publications are available in the appendices and on the IUTAM website.

# F. Celebrations

Looking back to the history of IUTAM there are anniversaries to be celebrated from time to time. Three of them are mentioned here.

## 1996: 50th Anniversary of IUTAM

In 1996 during the 19th ICTAM hold in Kyoto, Japan and presided by T. Tatsumi the 50th anniversary of IUTAM was celebrated. For this purpose all Congress participants received in their registration case a small booklet entitled "IUTAM, 1946–1996: Fifty Years of Impulse to Mechanics" published by Kluwer Academics. The author Fons Alkemade is a science writer from Amsterdam, The Netherlands. He got in contact to mechanics writing 1994 a biography on Jan Burgers included in the book "Selected Papers of J.M. Burgers" edited by F.T.M. Niewstadt and J.A. Steketee, and published by Kluwer Academic Publishers.

## 2000: Mechanics for a New Millennium

A unique event to celebrate the past and future of IUTAM was the 20th ICTAM hold in Chicago, USA. The subtitle of this Congress phrased by its President Hassan Aref was "Mechanics for a New Millennium".

The watercolour "Meters of Motion" by Champaign-Urbana artist Billy Morrow Jackson was the basis for the poster announcing the 20th International Congress of Theoretical and Applied Mechanics, held in Chicago in August 2000. The ICTAMs, initiated in 1924, have visited many of the great cities of the world: Copenhagen, Delft, Grenoble, Haifa, Istanbul, Kyoto, London, Moscow, Munich, Paris, Stockholm, Toronto, and Zürich among them. They are the "Olympic Games" of the science of mechanics, international forums where scientists and engineers from around the world meet every four years to present and discuss the latest results in the oldest of the physical sciences, the science of mechanics.

This beautiful watercolour tells the whole history of mechanics as a science starting with Archimedes, Galilei and Newton up to now as indicated by the team holding and standing on the IUTAM bridge.

Over the bridge to the right we find past officers of IUTAM: past presidents Sir James Lighthill, Paul Germain, Leen van Wijngaarden, Werner Schiehlen and Daniel C. Drucker. We also see past secretaries of IUTAM's Congress Committee, Niels Olhoff and H. Keith Moffatt. The president of the 19th Congress in Kyoto, Tomomasa Tatsumi, is in the middle of this group. Holding up the bridge are the President and Secretary-general of the 20th Congress, Hassan Aref and James Phillips of the University of Illinois at Urbana-Champaign.

More details are found in the book "Mechanics at the Turn of the Century" which was edited by W. Schiehlen and L. van Wijngaarden and published by Shaker Verlag.

## 2004: 80th Anniversary of ICTAM

The first International Congress of Applied Mechanics took place 1924 in Delft and this year is considered as the year of foundation of the ICTAMs long before the establishment of IUTAM. Chairmen of the first Congress were C.B. Biezeno and J. M. Burgers.

During the 21th ICTAM held in Warsaw, Poland the 80th anniversary was highlighted by the President of IUTAM, K. Moffatt from Cambridge, UK, featuring the close connection between ICTAM and IUTAM again. The subtitle of the congress was "Mechanics for the 21th Century" chosen by its President W. Gutkowski.

## 2016: 70th Anniversary of IUTAM

As already mentioned, IUTAM was created 1946 during the 6th Congress for Applied Mechanics held in Paris, France. The Statutes of IUTAM are still available today in French, and the 24th ICTAM takes place 2016 in Canada's French speaking province Quebec with the metropolis Montreal. The French title of ICTAM 2016 is also easily understood "La mécanique—fondement de la recherché multidisciplinaire". During ICTAM 2016 this second edition of the book "IUTAM —A Short History" published again by Springer will appear online with free access for everybody including all the participants of the Congress to celebrate the 70th anniversary.

# G. Prizes

The prizes awarded by IUTAM are closely related to the ICTAMs. There are two categories, the IUTAM Bureau prizes for young scientists and the Batchelor and Hill Prizes for scientists doing research on an emerging field or with a significant breakthrough within one full decade.

## IUTAM Bureau Prizes

Beginning with the 17th International Congress in Grenoble (1988), the Bureau of the International Union of Theoretical and Applied Mechanics has selected 2 or 3 outstanding young scientists for the Bureau Prize, based on their papers and their presentations at the Congress. To be eligible for the award, a presenter must be not older than 35 years of age at the time of the Congress, and must indicate that he/she wishes to be considered for the award. The award includes a certificate and a check of 500 USD. The recipients are announced during the closing session and on IUTAM's web page.


## Batchelor Prize

The Batchelor Prize, sponsored by the Journal of Fluid Mechanics, is presented once every four years at the ICTAM since the 22nd ICTAM in 2008. The prize winner is announced late in the year prior to the Congress.

The prize of $25,000 is awarded to a single scientist, for outstanding research in fluid dynamics. The research so recognised by the prize shall normally have been published during the ten year period prior to the announcement of the award.

The prize winner is determined by a small committee whose members are internationally distinguished experts in fluid mechanics. It is expected that the prize winner delivers a lecture at ICTAM and that this lecture will also be published in the Journal of Fluid Mechanics and be made freely available on the Cambridge University Press Journals website.


## Hill Prize

Elsevier has established a prize named The Rodney Hill Prize in Solid Mechanics. This prize, which consists of a plaque and a check for $25,000 is to be awarded in recognition of outstanding research in the field of solid mechanics. The prize is awarded every 4 years, to coincide with the quadrennial International Congress of Theoretical and Applied Mechanics (ICTAM).

An article based on the lecture shall be published in the Elsevier journal of the awardee's choice (as well as in the Congress Proceedings at the awardee's discretion), and shall be made freely available on the journal website.

The Prize winner is selected by a Selection Committee consisting of four members appointed by the IUTAM Bureau and constituted as follows: two chosen by Elsevier; one chosen by the Bureau of IUTAM; and one chosen by the Executive Committee of the Congress Committee (XCCC) of IUTAM.


# H. Concluding Remarks

During the past thirty years the principal objectives of IUTAM have proven again to be most valuable guidelines for its scientific leadership. The Union's objectives are based on the tradition of free international scientific cooperation, developed in the International Congresses for Theoretical and Applied Mechanics (ICTAM), and they shall observe the basic policy of non-discrimination and affirm the rights of scientists throughout the world to adhere to or to associate with international scientific activity regardless of race, religion, political philosophy, ethnic origin, citizenship, language or sex.

One of the reasons for the continuous success of IUTAM is the strong interplay between analytical methods and demanding applications in science and technology. Even if mechanics is sometimes described as a completed science within physical sciences it turns out that the multi-disciplinary approach of theory and applications is very attractive for young scientists fascinated by combining hard sciences with demanding applications.

The number of adhering organizations from countries and territories, respectively, was further increasing as well as the fields closely related to fluid mechanics, solid mechanics and dynamics. The rules of procedure represented by the Statutes have been steadily adapted to recent developments and new experiences for smoother cooperation. As a result personal contact and long lasting friendship often happens between scientists working within the fields represented by IUTAM. Therefore, ICTAM Congresses, IUTAM Symposia and Summer Schools are frequently experienced as reunions of scientists supporting their international relations and serving as worldwide scientific marketplaces.

From its foundation in Paris 1946 to the 24th ICTAM in Montreal 2016 IUTAM has shown a steady growth and is highly respected worldwide. It is expected that the spirit of cooperation will continue in the future and IUTAM will operate solely for the advancement of mechanics and the good of mankind.

# Congresses

Stephen Juhasz and Peter Eberhard

## A. Summary

During the past ninety-three years, one "initiating conference" and twenty-four international congresses—first under the name Applied Mechanics and later under the name Theoretical and Applied Mechanics (ICTAM) were held at about four-year intervals on four continents and in nineteen countries. The participants were primarily from academia, but there was always industrial support and some industrial participation. The number of countries represented by participants increased from seven in 1922 to fifty-seven in 2012. During the past ninety-three years, some 10,150 presentations were made in a multitude of formats. There were keynote and general lectures, "conversaziones" (research exhibits with refreshments), contributed papers, poster sessions, etc. Comparing the 1924 and 2012 Congresses, the number of participants increased more than sixfold and the number of papers presented nearly 22-fold. This does not give the full picture. Both participants and papers fluctuated greatly. The congress with the maximum number of participants was not the one with the maximum number of papers. The unusual 11-fold increase in participants (2250) at Moscow was due to the large size of the host country, intensive research activity, and consequently extraordinary number of local registrants. The number of participants from non-host countries varied between 440 and 1321 during the last 15 congresses. The total number of registrations was around 20,000 (number of individuals fewer because of repeated participations). The large increase of papers at Bruxelles (511) was due to a relaxed

S. Juhasz (1913–2013)
Southwest Research Institute, San Antonio, USA

P. Eberhard (*1966)
University of Stuttgart, Stuttgart, Germany

© The Author(s) 2016
P. Eberhard and S. Juhasz (eds.), *IUTAM*,
DOI 10.1007/978-3-319-31063-3_7

51

acceptance policy for contributed papers. Altogether up to the year 2008 thirty-eight volumes of proceedings have been published with a grand total of 16,748 pages and combined shelf length of over one meter, see the photos. The Beijing Congress 2012 was the first one where the papers were published only electronically as an e-book in the Procedia IUTAM series.

Although the history of mechanics goes back to Archimedes no fraternity of applied mechanicians existed until ICTAM. It was ICTAM that gave a feeling of kinship, in bringing about a framework for free exchanges and crossfertilization of ideas through face-to-face contacts as well as through publication of results. Actually ICTAM brought about a blossoming of applied mechanics, not unlike the Renaissance in arts and letters of the 15th and 16th Centuries. Without the contributions of 20th Century "mechanicians," current technology and living standards would be impossible.

On an international level, the accomplishments were just as great. After WWI, scientists who were natives of both Allied and Central Power nations convened. Likewise shortly after WWII, the scientists of the former enemy countries got together. During the then arising East-West differences, congresses were held regularly and scientists from capitalist, socialist, and third-world countries exchanged ideas freely. During the nearly ten decades an "esprits des mécaniciens" gradually developed. The international congresses were also the catalyst for the establishment of national congresses. There are several countries worldwide that regularly hold such national congresses. Also, regional series of meetings were initiated, such as Euromech or ECCOMAS Colloquia and various national mechanics congresses, held e.g. in Italy, Germany, Bulgaria, Poland, USA, Russia, China. Last, but not least, some thirty IUTAM symposia are held between two consecutive congresses. This would not have been possible without the nearly hundred years tradition of ICTAM/IUTAM.

The nature of the congresses changed. In the beginning it was the only medium for information exchanges in the broad field of mechanics. Increasingly, the congresses played an additional important role as a forum for people working in mechanics to meet. However, the published proceedings of the congresses contain a storehouse of composite knowledge, as well as some new material, thus the congresses served as a reference point to report and assess the progress that had been made.

# B. When and Where

## 1. Usual Cycles and Deviations

The original idea was to hold a congress every fourth year. This attempt had to be compromised three times. After the first congress in 1924, it was realized that some of the participants in the International Congress of Mathematics (ICM) would also have come to the ICTAM (International Congress of Applied Mechanics) if they had been held in different years. This recognition caused the two-year interval between Delft (1924) and Zurich (1926). The second deviation was caused by WWII. In 1938, in Cambridge, USA, during the V Congress, it was announced that the city designated for the VI Congress in 1942 was Paris. This was impossible, and the first opportunity to hold a congress in Paris was 1946. There was, thus, an eight year hiatus. This again caused a conflict with ICM. So another exception was made, and the VII ICTAM was held in London in 1948. There were thus two, two-year shortenings and one four-year expansion. Since the VII ICTAM Congress in London, all congresses have been on schedule as if the three "deviations" had not occurred.

The length of the congresses varied between five and eight days. Most of them started on a Sunday or Monday. The longest congress, held in Istanbul, lasted eight days. It started on Wednesday and included the weekend (to the delight of the participants). Quite a few later senior engineering scientists who were in Istanbul in 1952 were graduate students or young faculty members at the time. The majority of the congresses were held in late summer. There were two exceptions: the first Delft Congress was held on 17–26 April, and the Cambridge UK Congress was held between 3 and 9 July.

## 2. Countries and Cities

Most of the 24 congresses were held in the northern hemisphere, on two continents: Europe and North America. The organizers cannot be blamed for bias. All originators of mechanics congresses were Europeans, and before WWII, there was little significant mechanics research activity on other continents. In recent years this changed and congresses were held in Haifa/Israel, Adelaide/Australia, Beijing/China and Kyoto/Japan. This improves the geographic representation and involves more national mechanics communities. There were four countries which each hosted at least two congresses: the Netherlands, England, France, and the USA. (The locations of symposia, as distinct from congresses, became global already within the first two decades.) Many of the congresses were held in large cities, such as Stockholm, Paris, London, Moscow, or Beijing. Most of the others were held in university cities such as Cambridge, Stanford, Adelaide, and Lyngby. One congress was held in a resort, Stresa. The initiating conference, sometimes called the "0th Congress," was also held at a resort in Innsbruck.

It could be thought desirable to move the geographical locations of each congress as far as possible from the previous one, e.g. moving even between continents. However, maybe it will need some more time before we see the first ICTAM in Africa. Of the congresses held, there were only four early occasions when the location of a congress was less than 1000 km from that of the previous one.

## 3. Congress Organizations

The organization of congresses is nowadays the duty of two groups: the Congress Committee [and there especially the Executive Committee of the Congress Committee (XCCC)] and the Local Organization Committee (LOC). The duties of both groups are numerous, and the degree of success depends greatly on the thoroughness of the planning, which some feel should start as early as the end of the previous congress.

Of the organizers of the congresses, a few individuals should be named whose activity was extraordinary. Dr. von Karman's vision and the contribution of well-selected associates created the "zero-th" and the first congresses in Innsbruck and in Delft. These are described in more detail in the preceding contribution by G. Battimelli. The chief organizer and "martyr" of the VIII Congress, held in Istanbul, was E. Kerim who died a few months following the congress. Some details of this congress are outlined in "Socials."

Although the success of the Stanford congress in 1968 was due to several persons, N. Hoff, who has been active in ICTAM/IUTAM congresses since 1946, deserves the major credit. The preface of the proceedings of this congress contains the multilingual introductions of major organizers. Hoff's performance at the opening of the congress was a masterpiece of international diplomacy. To overcome the difficulty of providing a sufficiently large European participation at Stanford, the congress president raised, from U.S. government sources, contributions for a round trip charter flight from Amsterdam to Oakland. The entire organization of the transportation of some 150 scientists from Europe was in the hands of the secretary of the Congress Committee. This was the largest organized movement of people over a considerable distance in the history of these congresses.

Later, the Chicago Congress in 2000 was very impressive. Besides being held at an interesting place, the Congress President H. Aref spent many thoughts to create a program adequate to the millennium change. Actors took the stage to comment the history of mechanics and the reception was held in a beautiful museum.

Because the activity of the LOC is so multifaceted, the availability of a manual is most useful. The "little green books" prepared by the Toronto organizers contained job descriptions, both technical and social.

# C. Meeting Places, Residences and Transportation

## 1. Congress Locations and Facilities

The majority of the early congresses were held at universities. Here lecture rooms were usually the scene of contributed lectures, and auditoriums were used for general lectures and for opening and closing sessions. They were all indoors, with the exception of Stanford where the opening session was held in an excellent outdoor concert facility. Among the universities, the Lomonosov University in Moscow deserves special mention. This interesting "towering building" is the largest university building in the world, which permitted all lectures to be held under one roof, with but one exception: the opening session, which was held in a Kremlin auditorium.

There were several congresses where the meetings were not held at universities. The first was in Stockholm in 1930. Originally, the congress was supposed to be held at the Royal Institute of Technology. Due to a much higher preregistration than expected, the venue had to be changed. The lectures were held in the Swedish Parliament Building. The next congress not held at a university was in 1960, in Stresa, in the beautiful building of the Palace of Congresses. The third non-university building where a congress was held was the Congress Hall of the Deutsches Museum in Germany in 1964. This museum is one of the oldest technical museums in the world, and the proximity to the museum contributed greatly to the interest and the pleasant memories of the participants of the XI IUTAM Congress.

Later, because of the larger number of participants, more and more congresses were held in conference centers or big conference hotels. They offer usually great facilities intended for so many participants but on the other side the academic atmosphere of a university cannot be provided there. Certainly in the future less and less ICTAMs will be held at universities.

A very important place for the "one-to-one" or small-group scientific discussions were the coffee rooms. If any generalization can be made, the provisions for these was excellent at most congresses. Usually, large rooms with free coffee were available. At several congresses there was also provision to get pastry or sandwiches. Occasionally, there was also a piano, and there were frequently registrants who did not hide their musical talents. The lunches were usually cafeteria-style and quite fast.

## 2. Logos, Banners and Signs

It is not known when the use of "logos" started. Since the sixties, however, several congresses had a logo displayed, usually on a large horizontal banner, with a text such as "XV International Congress of Theoretical and Applied Mechanics." Such banners were usually placed on the front of the building where the registration was held. The logos were also displayed on programs, letterheads, nametags, etc. The Munich logo was the outline of the most famous church tower of that city. The Moscow Congress had two logos: the one on the nametag showed the outline of the Lomonosov University and textual information; the logo on the banner was circular, resembling a globe with the words "IUTAM" and "Moscow 1972" horizontally and "International Congress" on the circle. The logo of the Toronto Congress was more technically oriented. It was selected by the local arrangements committee from a number of submitted designs. They called it a "stylized version of the symbol for infinity." It is actually a two-dimensional representation of the "Moebius Strip". The circular logo of the Lyngby congress was interpreted by some participants as a ball bearing, by others as boiling water, or solid particles suspended in fluid. Actually, the local arrangers wished to represent the front end of a bronze horn used in Scandinavia some 3000 years ago. Two replicas of these horns were used at the opening ceremony. Later, logos often showed famous buildings like the Great Wall 2012, the Chicago skyline 2000, or local temples 1996. No previous logo was repeated. Although maybe not a "logo" but a symbol of a historical person, the picture of Euler was engraved on the registration folder given out in Zurich in 1926, with an implication that he was born and educated in Switzerland and was a scientist living and working in other countries and thus typified a great international scientist.

## 3. Residences and Meals

The participants usually stayed in hotels. As the character of the composition of participants changed when the number of delegates increased, so did the range of residences. At later congresses, first-class hotels, lower-rated hotels, and dormitories were used. At the earlier congresses, the participation of graduate students was an exception rather than general. They were usually housed in private homes.

The distance between the congress buildings and the nearest and farthest residence varied greatly. In Stresa, the longest distance was a 5-min walk. There was, however, an exception. Some latecomers such as G.I. Taylor were housed in a distant hotel. He was a good sport and once accepted a back seat on a motorcycle with M. van Dyke to get to his lecture. The other extreme was when the delegates had to take a train from the university city but were housed in different cities, which excluded an evening get together on nights without organized programs. There were complaints at several congresses about hotels, but there seems to be no correlation between their size and the price of accommodation and quality of service. An example of a congress where both the large hotels and small "albergos" were really excellent was Stresa. The question of residence is, of course, irrelevant for those who live in the congress city.

## 4. Transportation

At most congresses the transportation (where needed) was by special bus. There was usually no problem in the morning and at the end of the day, but during "unofficial" times the waiting was sometimes long. Transportation by rail and bus in Lyngby in 1984 was frequent and reliable, and most of the congress hotels were within walking distance of the interesting railway stations in Copenhagen. Nevertheless, the coming and going took nearly 2 h. The alternative, however, would have been to hold the meetings in Copenhagen in a convention center where the meeting room would not have been free (unlike the Technical University in Lyngby at that time); thus, the registration fee would have been much higher. In Munich, excellent street cars provided transportation through the beautiful city from the island on the river Isar to the railway station, near which most of the congress hotels were located.

## 5. Weather

Congress organizers are responsible for all aspects of the congress except the weather. Occasionally, some irate participants blamed even the weather on the organizers. It is worthwhile to name some good and some difficult weather conditions. The California weather permitted the outdoor opening ceremony and, on a later day, an outdoor dance at Stanford in 1968. The weather, not always reliable in England even in the summer, was exceptionally fine throughout the whole 1934 Cambridge UK Congress and made the evening functions on the college grounds particularly enjoyable. The weather in Istanbul was over 38 °C on the day of G.I. Taylor's keynote speech. Both the speaker and the audience had ties and jackets on. Sir Geoffrey quickly noticed the suffering of the audience, and removed his tie and jacket which "broke the ice" and the audience followed suit. Similar hot weather, with a forest fire 100 km from Moscow, caused difficult conditions during the XIII Congress in 1972. This was an unusual situation.

# D. Participants

## 1. Number of Participants

The number of participants, at each congress as shown in the appendix, was extracted from the published itemized list of participants in the proceedings or from the IUTAM reports. The lowest number was 207 in Delft in 1924, and the highest number was 2250 in Moscow in 1972. In the post-WWII period, there was a great decrease between 1972 and 1976. In Delft, the participation was 1005. The major reason for the decrease was that more than half of the Moscow participants were from the USSR, while only seven participants came from the USSR to Delft. The Toronto registration was also lower than expected, mostly due to a decrease in European participation because of high travel expenditures. Later in Chicago (1430) and Warsaw (1515) the numbers increased greatly again. The number of out-of-country participants fluctuated between 440 and 1321 during the last 15 congresses and their percentage between 34 and 95 %.

## 2. Number of Host Country Participants

It is obvious that the fraction of registrants from the host country will always be considerably larger than usual. A congress held near home is an excellent opportunity for young engineers and scientists to hear presentations by the great mechanicians, and perhaps even to shake their hands and hear their jokes at lectures or functions. We computed the fraction of the number of locals. The lowest host country participation, as shown in the appendices, was 5 % in Bruxelles. This is not surprising as the population of Belgium is one of the smallest of the 19 countries where a congress was held. The other extremes were the USA (72 and 66 %) and in the USSR (63 %).

Stanford and Moscow and later Delft, Chicago, Warsaw, Adelaide, and Beijing had over 1000 registrants. Such a large attendance has both favorable and less favorable aspects. It should be recognized that if a congress is held in a large, technologically developed country having many research engineers and scientists, the "Locals" inevitably decreases the international flavor compared to a congress that is held in a smaller country.

## 3. Travel Grants and Other Support

Already before WWII, it was recognized that some graduate students (or researchers) should be included among participants. Travel grants were issued by host organizations, such as by a college in Cambridge, UK and the Royal Society. Later, IUTAM gave support again and again. Some governments, for example the Turkish Government, gave support such as free dormitory space. About 5–10 % of the participants of some congresses came with such a travel grant as here mentioned. An unusual indirect travel grant, as mentioned earlier, was given to about 150 Europeans who came to Stanford and returned by chartered plane from the Netherlands paid for by different US agencies.

## 4. Founders Participation

T. von Karman participated consecutively at 1 + 10 conference and congresses. J. Burgers participated in nearly all congresses starting in 1922. His last participation coincided with the second Delft congress in 1976. G.I. Taylor was recognized in Stanford as having presented at least one paper at all congresses from I through XII. Incidentally, this was his last participation. C.B. Biezeno participated at I–XI and, though not personally present, sent a greeting cable to the XII Congress.

## 5. Nature of Participation

If the list of registrants of ICTAM/IUTAM is perused and compared with group pictures of I and VII Congresses (as shown in the appendix), it is interesting to see the gradual decrease in the age of persons who attend the congresses. Originally, only recognized leaders in the field participated, see General and Sectional Lecturers: I Delft 1924 in the appendix. The reason behind this gradual change is twofold. One is that in the beginning the congresses were the outlet for scientific research, and publication was only for the "in people." The other reason is that, since 1934 (Cambridge UK), the International Congress Committee agreed that:

... the Congress should not be viewed as a publishing body but rather as a meeting of
people; papers therefore should not be accepted from members who have no intention of
attending the congress unless adequate steps are taken for the reading of the full paper by
someone conversant with the author's work ...

# 6. List of Participants and Name Tags

If a person comes to a congress, it is important for him or her to know who else is
there and who are the persons from a given country. Also, it is useful to know the
affiliations. Of course, it would be highly desirable to have the information at the
beginning of the meeting. Two tools are used to accomplish these purposes: the list
of participants and name tags.

At the Innsbruck conference, with some thirty participants (including family
members), the solution was trivial. Everybody wrote by hand the name, city, and
hotel name. A scan of the original list is included in G. Battimellis chapter. At a
meeting with over 1000 persons, this of course is out of the question.

An excellent system was used in Lyngby during the XVI Congress. At the time
of the registration, three computer-prepared lists were supplied that contained all the
information given above, with the exception of affiliation and address. There was an
alphabetic listing of all names (family name first) followed by country; then there
was an alphabetic listing by countries with subalphabetization of the names; finally,
a listing by hotels again subalphabetized by family names of participants. This list,
which was available at registration, included all persons who preregistered. A few
days later, a supplementary list was published. It is worth mentioning that the III
Congress (593 registrants), held in Stockholm, had a program in which the pre-
registrants' names, addresses, and affiliations were given with photos of themselves
and spouses if present. It was printed on a glossy paper.

The name tag and its usefulness or uselessness depends greatly on the read-
ability. There were several congresses with excellent readable name tags, such as
the XV Congress in Toronto. One of the reasons for unreadable or difficult to read
name tags is the use of regular too small letters to accommodate fully the given
name, family name with country, and affiliation.

With increasing sensitivity with respect to providing personal data, however,
many participants would nowadays not like to provide info about private address or
used hotels.

# 7. Photos

Group photos of congresses were feasible when the number of participants was
small. Photos of some early congresses are available. The last group photo was

taken in London in 1948. Of course, making group photos of the entire set of participants is no longer feasible.

At later congresses, in lieu of group photos, a multitude of photos were taken by the official congress photographer, who came to coffee breaks, socials, and excursions. These were displayed a few hours later in the registration area where they could be ordered. Examples of interesting group pictures taken by an official photographer show von Karman with the Japanese delegation on the steps of the Université Libre in Bruxelles in 1956, and Academician Sedov with von Karman on an excursion boat on the Lago di Maggiore in 1960.

Alas, there were several congresses where there was no official photographer. Even in these cases, pictures were taken by some participants, but these were not displayed in time; also, they could not be ordered. The photos displayed in the last part of this book are often taken by official photographers and by congress participants.

# E. Presentations

## 1. Numbers

Due to increased scanning, the number of papers presented has first not increased as in some other scientific fields. As a matter of fact, the number of papers at the first congress was 58, and the number of papers in the post-WWII Congresses was around 300, with the exception of IX Bruxelles (511). The number of papers rose to over 1000 at XX Chicago and the following congresses. This was partially due to the fact that most participants do not receive travel support from their home institutions if they do not present a talk or poster.

## 2. Formats

In Innsbruck, there were just "lectures." This expanded in I Delft to general and sectional lectures. Then at IV Cambridge, UK (1934), the format was expanded and contributed papers were accepted. Since then, the bulk of the papers presented were contributed papers. At one congress, some of the contributed papers were given more presentation time than others. The contributed papers belonging to the first category were published, and papers belonging to the second category were just listed by title. During the last congresses, some of the contributed papers were presented, but others were displayed and discussed at poster sessions. This is always causing a lot of discussions and some people consider posters as a 'second-class' presentation.

During the past congresses, during the contributed, presented-paper sessions, "special emphasis" sessions were held. Superficially, they appear to be symposia embedded into congresses. This was, however, not the intention of the arrangers. For example, at Lyngby in 1984, there were three such topics: Marine Structure, Wave Interaction, Micromechanics of Multicomponent Media, and Development of Chaotic Behavior in Dynamic Systems. The original purpose of these sessions was to bring persons of different disciplines together.

Poster sessions can be very useful and are the "children" of the "conversazione" type meetings held at the 1934 and 1938 congresses. The poster papers were treated differently from the contributed papers, not only with regard to mode of presentation but also of publication. Poster papers were listed only by title in alphabetical order of the author.

The list of most authors of keynote and/or general lectures, with congress designation, is given in the appendices. Another appendix lists the general and sectional papers presented at the first congress.

# 3. Acceptance Policies for Submitted Papers

Although keynote, general, and sectional papers were always invited papers, contributed papers, as the name indicates, were usually unsolicited. They went through some screenings by methods varying in severity, but it appears that the activity of the members of the Papers Committee (later the Congress Committee and again later the International Papers Committee) consisted of only a go/no-go decision, rather than a traditional refereeing.

There are no records about policies of screening. It is known that before the V Cambridge USA, 1938 Congress, a program committee was established composed of S. Timoshenko, H. Dryden, J. den Hartog, and H. Peters, who read the abstracts of the contributed papers and selected those to be presented. It appears that the high number of papers in Bruxelles was due to the fact that nearly everything was accepted. The chaotic result of the acceptance of virtually all submitted papers at the Brussels Congress led the Congress Committee to establish a Papers Committee (consisting of M. Roy, E. Becker and W. Koiter for the Stresa Congress). In the screening there are two considerations which sometimes conflict. One is the quality of the paper, and the other a striving for wide international participation without dominance of a few countries. The paper committees of both XII Stanford and XIII Moscow were much stricter than that of IX Bruxelles. Only one of five of the contributed papers was accepted for Stanford and one of six for Moscow. At Stanford the ratio of presentations to participants was 21 % and at Moscow 11 %, as compared to the average figure of all congresses of 44 %.

The task of the International Papers Committee is far from easy and it has been always assisted by a pre-selection of papers in the major countries by their national committees. Responsibility for the final decision has always remained with the International Papers Committee.

During a closing session of one of the congresses, the chairman of the LOC asserted that the slides used in the lectures were poor. This had a beneficial effect in two future congresses (XV and XVI). Potential authors were advised that the acceptance (or rejection) of papers would be based not only on manuscript of the papers but also on the drafts of the slides. Later this strategy was changed again and the evaluation in the last decades was based on a short abstract.

## 4. Languages

Each congress has "official languages." At the early congresses, German, French, and Italian were the major languages, and English was the exception. Nowadays, the reverse is true. As a matter of fact, English is used sometimes by participants who have difficulty even reading English. At a recent congress, a delegate from a non-English-speaking country read his paper. It took quite a time for the audience to realize that he was not reading the paper in his native language but in English. Incidentally, there was (as far as is known) never any simultaneous or consecutive translation. Actually, practically everybody understands spoken and broken English.

# F. Non-Lecture Programs

## 1. Socials, General

As ICTAM/IUTAM really became a "meeting place" rather than a publication outlet, one might therefore assume (not knowing what happened) that nowadays more emphasis is placed on socials by the LOC's than at the beginning. Actually, the situation is the reverse if the emphasis on socials is measured by number of nights without any social program. During the pre-WWII congresses, the number of "un-programmed" nights was one or two. After WWII, there were usually three or four nights without any socials or other programs, and the participants, or at least the bulk of the participants, were left on their own. There were, however, two post-WWII congresses, the socials of which were exemplary, both in numbers and also in the imagination of the local executive committees. One was the Istanbul congress, which was planned by E. Kerim. Below, we cite from the paper "Notes on the Eighth International Congress on Theoretical and Applied Mechanics, 1952 Istanbul, Turkey" by D. Drucker and E.H. Lee published in 1952 (see references).

> The great friendliness and sincere interest displayed by the visiting members were, however, completely overshadowed by the truly remarkable hospitality of our Turkish hosts. It is impossible to think of anything additional they could have done in organizing the congress. There were truly impressive receptions by the Rector of the University of

Istanbul, by the Rector of Technical University of Istanbul, and by the Minister of Public Education. The banquet at the Municipal Casino combined fine food and entertainment. A Sunday boat trip on the Bosporus enabled members to obtain the best views of the natural beauty of Istanbul and its surroundings. An evening of authentic folk dances and songs presented in an outdoor amphitheater gave the flavor of an older culture now replaced by a modern one. All these, and there were more, represented but a small part of the work of the Organizing Committee. Buses were always available for transporting members to places to which they wished to go; guides and information personnel were always ready to help. Our thoughtful hosts also left enough free time for individual sightseeing, shopping in the bazaar, visiting mosques and museums, and for all the other fascinating diversions afforded by the cosmopolitan city of Istanbul.

The other post-WWII congress, which is noted and recalled again and again, was the XII Congress held in 1968.

## 2. Topical Description of Past Socials

From the point of view of meeting other participants and seeing who is present, those socials that include mixers (whether alcoholic beverages are served or not) are the most beneficial. Such affairs are "early bird" parties, banquets, museum visits with receptions, etc. Boat excursions fall in the same category (at least four times during the congresses), and they were most popular. The personal touch in planning is the most important element and is more essential than the food served. At one congress the outgoing university president gave a beautiful garden party, although most of his furniture was already crated. Everyone felt that the hosts really cared.

In city tours, plant visits, and theaters, social contact is limited to speaking with a few participants. It appears that 15–25 % of the participants arrived with spouses and/or other family members. In general, these spouses were well taken care of during the days the participants had their technical sessions.

## G. Exhibits

### 1. Conversazione

The "Conversazione" is an Italian word meaning "conversation" or "evening party." It has been used to describe technical exhibitions, combined with refreshments in two ICTAM congresses in 1934 in Cambridge UK and, in 1938, in Cambridge USA. Details of the first one are not available. G. Batchelor mentioned that the Royal Society of England has used this mechanism and name for a considerable time.

Some details of the Cambridge (USA) Conversazione in 1938 are given in the appendix. Basically, it was an evening "session" in which a multitude of professors and some mechanics experts in the East Coast area demonstrated their experimental setup in the MIT mechanical engineering department laboratories. The list included in the appendix reads like the Who's Who in Mechanics in the middle of the century. Their topics make fascinating reading.

It seems that this mode of communication at IUTAM congresses did not continue. This was due not only to the eight-year time lapse between Cambridge and Paris, but, also, because it was impossible to repeat it in war-torn Paris in 1946. Actually, conversazione continued in two different ways, even if not at IUTAM. The "science fairs" are "communicaziones" in much less sophisticated ways. Also, the poster sessions used nowadays worldwide have some relation to conversazione, though the latter is more discussion than equipment oriented.

## 2. Publisher Exhibits

Book and journal publishers had exhibits at most of the congresses since the IX Congress held in 1956. The number of publishers who exhibited were from the world's leading ten or so technical and scientific publishers. Occasionally, book dealers exhibited as in Munich and also in Delft in 1976. All these exhibits were usually quite simple; frequently, even the publishers' names were not displayed. The best and largest book exhibit was in Toronto where more than five book publishers exhibited. They were both from the US or from Europe, or publishers with operations on both continents. These exhibits showed books and also journals. Occasionally a handout list of the exhibited books was available, with the name, time, and location of the exhibit. It is clear that the publishers did not put much effort into "jazzing up" the exhibits at IUTAM as they have always done, e.g., at the Frankfurt International Bookfair, where the number of visitors is a few orders of magnitude higher.

## 3. Institutional Exhibits

Four institutional exhibits have been organized by Stephen Juhasz in cooperation with the Local Organizing Committee since 1952 as follows:

| Congress | | | Exhibit theme |
|---|---|---|---|
| No. | City | Year | |
| XIII | Moscow | 1972 | Famous Mechanics Scientists |
| XIV | Delft | 1976 | From Delft to Delft |
| XV | Toronto | 1980 | Mechanics in Action |
| XVI | Lyngby | 1984 | Short History of IUTAM |

The first three of these were composed of a 10 m long and 4 m high "histowall" with an inclined table. The fourth was of similar design but 5 m long. Unfortunately these exhibits are not available anymore.

# H. Proceedings

## 1. General

The ICTAM/IUTAM Congress Proceedings are the main permanent record of the congress activities. The organizers should be commended for the fact that all proceedings of the congresses, with the exception of the VI Congress, were published. The proceedings were published by commercial publishers and/or by universities where the congresses were held. The proceedings of the XXIII Congress in Beijing were the first ones only published electronically. They were available as 'open access' which means that everybody may download them from the internet and read them for free.

Nowhere in the world is there a place where all published proceedings are under one roof. Efforts were made by S. Juhasz to compile, temporarily, a full set up to the XV proceedings. For this, copies were borrowed from five other libraries. Published proceedings of the ICTAMs 1924–1980 were collected on loan from six countries while it was much easier to collect the proceedings of the ICTAMs 1984–2008. The proceedings of the ICTAM 2012 were only published electronically. The photos show pictures of the full set. The bibliographic details of the Innsbruck Conference and of all published proceedings are listed in the appendix.

## 2. Publication Modes

In general, the ICTAM/IUTAM undertakings were of a high standard. The quality of the proceedings is, however, uneven and incomplete. One can explain the situation. There was no permanent ICTAM headquarters, and even now the IUTAM secretariat changes locations. Furthermore, the publication of the congress proceedings is not the duty of the general secretary but of the local organizers.

Two charts in the appendices give details of the proceedings. The first lists the congresses with the number of pages, papers and the mode of publication (papers given fully or by abstract or listed only by title). The other chart analyzes in generic terms the published proceedings.

## 3. Tables of Contents

Although the tables of contents are part of the proceedings, they are discussed here briefly because of their importance.

One of the congress proceedings had no table of contents, and only the names of the general lecturers were given, without the titles of their papers. In the same proceedings, the contributed papers were given by abstract, and several hundred pages had to be scanned to see what the volume contained. In other proceedings, the locations of the tables of contents varied, also, and so did the sequencing of papers and the elements in the tables of contents. Some proceedings had the table of contents divided; the keynote, general, and sectional papers were listed at the beginning, and the list of contributed papers at the end.


# I. Accomplishments

## 1. Advancement of the Art

After discussing with a multitude of mechanicians the accomplishments of ICTAM/IUTAM—or more specifically how it advanced the art—one gets the definite feeling that the activities in 1922 and 1924 were the beginning of a blossoming of applied mechanics not unlike the Renaissance in arts and letters of the 15th and 16th Centuries.

J. Lighthills contribution at the beginning of this volume indicates the view from a perspective of 60-plus years. To show the same from a perspective of 14 years after the start, we reproduce below a 1938 statement by K. Compton, physicist and former President of MIT. This was part of his opening address at the Vth IUTAM Congress held in Cambridge USA.

> An intelligent person who is not particularly acquainted with the field of applied mechanics might well ask the questions: 'Why would there be this worldwide interest in applied mechanics? Surely mechanics is the oldest branch of pure and applied science, and its principles have been well established for many, many years. Did not Archimedes discover the principles of statics and hydrostatics and Galileo the laws of motion; and did not Isaac Newton formulate the basic principles of dynamics? Was it not Lagrange who, in his famous equation, stated the laws of mechanics in a generalized, yet usable form? With the establishment of these fundamental laws, refined for certain purposes by the principles of least action, and Hamilton's principle, and with the mechanical theory of heat well-established in those principles of thermodynamics which were developed a generation or two ago, what is there left to attract the serious attention of an international body like this?

Compton continues:

> To answer these questions, there is a vast difference between the establishment of a fundamental principle and its application to specific problems, for the principle may be simple and the application very difficult. From my own field of physics, there is an interesting example of this difference. About a decade ago, there was great activity among theoretical physicists in developing the principles of quantum mechanics, having to do with the application of mechanics to atomic structure and radiation. In reporting one of the most brilliant developments of this subject, the able young physicist who was responsible for it wrote in the opening paragraph of his paper as follows: 'Now that the quantum mechanics has given us the explanation of all of chemistry and most of physics, etc., etc.' Yet, as a matter of fact, this development of quantum mechanics has been applied precisely only to hydrogen, of all the chemical elements, and to no molecule and to no matter in bulk. The reason is that the application becomes so complicated in systems more complex than the hydrogen atom that, at best, only approximations to the theory can be made.

E.H. Lee now comments on an expanded role of mechanics in the area of constitutive theory which forms an adjunct to the "basic laws":

> It seems to me that an important part of the development of Applied Mechanics has been concerned with the formulation of constitutive relations to express force-deformation properties of materials, i.e., the basis for valid stress and deformation analysis beyond the yield (plasticity), plastics (polymer fluids and solids), rubber elasticity and even heavy gases. This supplies the basic physical information for nonlinear finite deformation continuum mechanics. Such questions play a role in current research and fall outside *applying the fundamental principles of mechanics.*

## 2. Building Contacts Between Participants

At the earlier congresses, as has been stated, the number of participants was low and most people knew each other or knew about each other through their publications. The situation changed continuously. Eventually, more people met at a meeting who never heard from each other and who, (as turned out occasionally during the discussion of a contributed paper) were going through the same agony in trying to solve identical problems, but were using entirely different approaches. These contacts led to successful life-long correspondence, mutual visits, and friendships.

Another type of personal contact can occur between a young graduate student and the "great mechanician." A now internationally recognized Stanford mechanician was in Istanbul. He said that although he had not formally met J. von Neumann, he shared the same handle on the crowded street car. Another interesting story on personal contact is mentioned in the preface of an early congress. The participant, seeing the name tag of Professor Milne-Thomson exclaimed, "I knew your writing well, but always believed that you are two persons."

## 3. International Aspects

We should consider not only person-to-person relations but also the nation-versus-IUTAM ones and vice versa. It is felt that one of the great accomplishments of the applied mechanics congresses was to build bridges. Until to the 40's, the bridges to be built were between scientists from Central Powers and the Allies. After WWII, the bridges to be built were between mechanicians of the Allies and the defeated Axis powers and later between the scientists of East and West.

ICTAM/IUTAM was fabulously successful in all three periods. Many friendships were formed after sitting across the table at a banquet or sitting side by side on the excursion bus. One of the participants in Munich described that academician L. Sedov embraced S. Timoshenko, whom he had seen for the first time after many decades. A member of the General Assembly, then a graduate student, was running after an IUTAM bus in Istanbul. At the lecture he found out the other "graduate student" who did the same was the young professor, M.J. Lighthill. Another interesting East-West cooperative was the large "Famous Mechanics Scientist Exhibit," which was jointly executed by a critical review journal in the U.S. and its USSR counterpart, the Referativnyi Zhurnal Mekhanika with the cooperation of G. Mikhailov of the Lomonosov University. G. Mikhailov was also the chairman of the local organizing committee of the XIII Congress in Moscow. At a social during the Zurich congress, one of history's great mathematics mechanician, Euler, was celebrated as a person who obtained his education in Switzerland and performed much of his scientific work in the USSR and other foreign countries.

There was considerable mixing between delegates of all socialist country participants and other participants. One participant of the Toronto Congress (1980) commented that there were even cordial exchanges between the Russian delegates and recent emigrants from the USSR.

What was really accomplished was the creation of an "International Corps of Mechanicians," which, while somewhat of the nature of an "Invisible College," becomes visible every fourth year.

# Symposia

Stephen Juhasz and Peter Eberhard

## A. Summary

Between 1949 and 1985 a total of 139 symposia were held in 91 cities in 25 countries. The frequency increased from one per year to about eight per year. From 1986 to 2015 239 symposia were held in 135 cities and 37 countries. This means that 1949 to 2015 in total the impressive number of 378 symposia were held in 41 counties. The total number of registrants was about 29,000. The lowest number of registrants was 23 and the highest 239. The greatest number of countries represented at a symposium was 27.

Most of the symposia publish proceedings. The majority of the published proceedings are in book form (281 of 378, i.e. 75 %). A smaller fraction (62, i.e. 16 %) were published in a special issue of a scientific journal or serial title. Only 10 symposia (2.6 %) had no proceedings at all and 9 are not yet available. Although just started very recently, the Procedia IUTAM series contains already 16 (4.2 %) proceedings. The symposia, unlike the congresses, were meetings of specialists and by invitation only, although in recent years these have frequently been just a formality.

S. Juhasz (1913–2013)
Southwest Research Institute, San Antonio, USA

P. Eberhard (*1966)
University of Stuttgart, Stuttgart, Germany

© The Author(s) 2016
P. Eberhard and S. Juhasz (eds.), *IUTAM*,
DOI 10.1007/978-3-319-31063-3_8

# B. Origin

## 1. Concept

The word *symposium*, according to the *Oxford English Dictionary, A New English Dictionary* (Clarendon Press, 1970), has two meanings: (1) "a drinking party: a convivial meeting for drinking, conversation, and intellectual entertainment: properly among the ancient Greeks…" and (2) "a meeting or conference for discussion of some subject or; hence, a collection of opinions delivered, or a series of articles contributed by a number of persons on some special topic." The latter meaning arose indirectly out of the former from the celebrity of Plato's famous record of the discussion (and the drinking) at a remarkable 'symposium' in which Socrates played a leading part.

It appears that the first symposium discussing a scientific subject was held far more than 100 years ago, "… for the liberal discussion of the Copernican system." Since then, the word *symposium* has been used increasingly to describe important scientific discussion meetings.

## 2. IUTAM Symposia

The predecessor of the IUTAM Congress was the Innsbruck Conference on Applied Mechanics, held in 1922. The conference was organized mostly on the initiative of Theodore von Karman, who envisioned bringing together in a hotel in Innsbruck a small group of specialists in the mechanics of fluids. The number of participants was under thirty, and as the meeting was held in a hotel, there were no "outside, local" participants. The majority of the participants had a family guest. This meeting was, however, not called a symposium. Nearly 30 years later, when von Karman was the chairman of AGARD in Paris, he initiated the first IUTAM Symposium in 1949. There, after 25 years of the quadrennial congress, the original concept of the symposium was introduced to allow the newly formed International Union to be much more continuously active than quadrennial congresses could permit.

# C. Frequency

During their first ten years (1949–1959), symposia were relatively infrequent, about five being held during each four-year interval between congresses. Between 1960 and 1972, however, the number of symposia held between congresses was at a much higher level, namely twelve. The number of symposia then increased: between 1972 and 1978, to four or five per annum, and thereafter, as a result of a new policy firmly adopted by the General Assembly, to an average of eight symposia yearly. During the three-year period 1982 to 1984, as many symposia were held as had been held during the 14-year period from 1949 to 1963. The highest number of symposia held in one year was 14 in 1997. It appears that interest in organizing and participating in symposia was increasing, and the main limiting factors are finances and quality of proposals. In recent years the situation changed somehow since the General Assembly receives less proposals and the competition with other scientific socities which organize conference series is increasing. However, still only high-quality proposals are accepted which promise interesting discussions about challenging fields of mechanics.

# D. Participation

## 1. Policies

The early meetings were essentially closed, similar to the Innsbruck conference in 1922. This policy seems to have held for quite a while, and young scientists who had not demonstrated significant activity in the field were not invited, even if they expressed interest and were locally recognized. The Bureau of IUTAM was concerned to ensure a high level of excellence for the discussion through a policy of attendance by invitation only.

Nowadays, the symposia have opened up. Two factors have contributed to this trend: first, though symposium participation is by invitation, promising young applied mechanicians, whose professors are recognized authorities, stand a good chance of being invited to participate on their professors' recommendations. The second is the greater participation of the "locals." This means that if a meeting is held at a university, then the junior faculty and graduate students may be present at the meeting. Discussions with IUTAM and symposium participants indicate that the trend is toward further relaxing the closed atmosphere of symposia and today participation is not restrictive anymore. More limiting is the maximum number of talks which can be admitted without introducing parallel sessions.

## 2. Fully Open, Fully Closed, and Middle-of-the-Road Meetings

A fully open symposium has the disadvantage that a "concentration" of experts is absent and, therefore, a high-level discussion would be impossible. A fully closed meeting, on the other hand, would create bad blood and would, in fact, create an "invisible college." The current middle-of-the-road kind of system may be able to balance the two extremes satisfactorily. An added plus to such a system is the possibility of giving financial aid to some bright newcomers who are geographically remote from the location of the symposium. Since many years IUTAM is offering grants to the organizers to support participation of young scientists who can otherwise not attend.

# E. What and Where

## 1. Subjects and Approval of Proposals

A subject heading list was used to classify the subjects of the past 378 symposia held between 1949 and 2015.

| | | |
|---|---|---|
| Foundations and basic methods in mechanics | M | 20 |
| Vibration, dynamics, control | D | 53 |
| Mechanics of solids | S | 95 |
| Mechanics of fluids | F | 121 |
| Thermal sciences | T | 9 |
| Earth sciences | G | 10 |
| Energy systems and environment | E | 6 |
| Biosciences | B | 10 |
| Other | O | 8 |
| Fluid structure interaction | FS | 3 |
| Materials | MA | 16 |
| Micro-Nanomechanics | MN | 12 |
| Computational methods | CM | 11 |
| Optimization | OP | 5 |
| Total | | 378 |

The majority of the symposia have been discipline-oriented. Some of the subjects of the symposia have been quite broad, such as Optical Methods in Mechanics of Solids, and others have been very narrow, such as Behavior of Dense Media Under High Dynamic Pressures. There has not been a drastic change in emphasis. The emphasis on cosmical gas dynamics, which was great during the first twenty-four symposia, has disappeared. There has been practically no change of emphasis on dynamics, mechanics of solids, and mechanics of fluids. A typical mission-oriented symposium was the 1975 symposium on Dynamics of Vehicles on Roads and Railway Tracks. Continuation symposia have appeared a few times; Creep in Structures symposia have been held every ten years since 1960. Although there is some justification to repeat a symposium a few years later because of the rapid developments in the field, the disadvantage of this practice is that as the number of symposia is limited, e.g., by financial resources, some fields not previously covered might be neglected or excluded.

Initiation of a possible symposium is usually by one or several persons from the same institution. This is brought to the attention of the Bureau of IUTAM or using a webpage to the attention of the Secretary-General, which in turn, submit the proposal for prescreening to one of two panels. One deals with mechanics of solids only, the other with fluid mechanics. Proposals dealing with both, Fluids and Solides, are evaluated by both panels. The panels then meet in person or discuss by email, and their recommendation is submitted finally to the General Assembly. The committees classify the subjects of the proposals as alpha, beta, and gamma type. The first one is affirmative recommendation, the third is unconditional rejection, and the second is an intermediate designation. Recommendations are made every second year for discussions by the General Assembly. These meetings usually last two days. The first day, recommendations are made. There might also be a discussion during the first day. The final decisions are made on the second day. Usually, only a small fraction of the recommendations by the panels are changed by the General Assembly.

The locations, scientific committees, and time, though already recommended by the proposer and possibly discussed by the committee and/or General Assembly, are worked out in detail by the Bureau of IUTAM. In the past, this was purely the prerogative of the Bureau, but this was gradually changing and the organizers prepare all these decisions today.

It might be of interest to look into fields in which there is worldwide activity that have never been proposed yet are still in the area of applied mechanics, though there is not a heavy enough concentration that might trigger the submission of a proposal.

It occasionally happens that a proposal is made by an individual who is not adequately experienced in the field and thus does not have the necessary standing or who may not be capable of organizing a good meeting, though the proposed subject is a popular one. In such rare cases, another initiator is found, and to show appreciation for the original initiator's effort, he is made a co-chairman, or some other compensatory solution is found.

## 2. Locations

IUTAM symposia have been held on each continent at least once. 41 is the total number of countries where one or more symposia were held. The greatest number have been in Europe, where a total of 241 out of 378 symposia were held. The second highest concentration is in America with 67 and then Asia with 57, Australia with 10 and Africa with 3. Strangely, symposia have never been held in some technically developed countries, such as Argentina, Korea, Iran, or Finland.

# F. Participants

## 1. Geographic Breakdown

It might also be worth looking at symposium locations from the point of view of whether they have been in larger cities or resorts. Examples of resorts are 1950, Pallanza, Italy; 1962, Celerina, Switzerland; 1972, Bad Herrenalb, Germany.

The lowest number of countries represented at a symposium occurred in Bangalore/India and Paris/France in 2014, when delegates from only 4 countries participated. The highest number of countries represented was 27 in 1972 at Karlsruhe/Germany. There were two reasons for this: (1) it was jointly sponsored by IUTAM and the International Association of Hydraulics Research, and (2) European countries are relatively small, close together, thus travel is easy. Based on 368 symposia where the participants' countries were listed, the average number of countries represented at a symposium was about 14. This, of course, is high enough to lend an international flavor to the meeting.

## 2. Numerical Breakdown

The figures in this part of the study were taken from the IUTAM reports rather than from the proceedings. Thus, it is not known whether the figures include some "locals" or not; it is, however, assumed that in the majority of the cases, they do not. The lowest number of listed participants was 23. The highest number was 239 in 1979 at Karlsruhe. The difference is due to the topic, the attractivity of the place, the reputation of the organizers but also due to the policy of the organizers. Some are very restrictive in sending out invitations; others have more an "open house" style. In total about 29,000 participants have been reported which means that in average about 76 participants attended a specific symposium.

The pattern of participation throughout the years is so random that no trend can be forecast. A smaller meeting has certain advantages. Also, it should be remembered that some of the larger symposia (in terms of the number of participants) are identical to many international congresses held prior to the 1950s.

# G. Proceedings

## 1. General

The symposia proceedings mainly fall into two categories: books and special issues of journals. It is interesting to note that the number of symposia published in book form is about 4.6 times larger than the number published in journals. There have been a total of 359 symposia published, 10 symposia had no proceedings and 9 of the very recent symposia are still in the process. In total, 74 % were in book form, 16 % were special issues of a journal, 4.2 % are in the just recently established Procedia IUTAM series, and just 2.6 % had no proceedings at all. Several publishers with high reputation have been responsible for most of the published books, but clearly Springer/Kluwer have published the largest share of symposia proceedings. Recently the open access publication in the Procedia IUTAM series by Elsevier attracted a lot of interest. It is also visible, that there are more and more symposia with no proceedings at all although IUTAMS's policy and rules demand proceedings. This causes some discussion in IUTAM and it is currently unclear what will be the result for the future: mandatory proceedings, voluntary proceedings, or no proceedings.

## 2. Advantages and Disadvantages of Publication Modes

If one considers the advantages and disadvantages of publishing the papers and possible discussions presented at a symposium in book form as compared with a special issue of a journal, then special consideration must be given to the four types of users to whom the mode of publication is of interest. These are the author of the paper, the applied mechanician at large (who might or might not be a specialist in the field but who is, for one reason or another, interested in the subject); the organizer and/or organization behind the symposium; and the librarian, who is a potential buyer of the book and subscriber (or not) to the journal.

For the organizers, the book form is advantageous for several reasons. They can keep much better control of the publication, even though another entity (the publisher) issues the book. Author prepared manuscripts are used in most cases, as the responsibility for following the guidelines, such as availability of abstracts, presentation of equations, etc., is usually the author's, and it is highly likely that the author will pay much more attention to the details than he would to preparation of a journal article, where the author receives only galley proofs. Also, the organizers receive a reasonable number of free copies of a book for distribution. For the author of a paper presented at a symposium, the publication in book form has several disadvantages: referencing a symposium paper is more difficult; also, in a list of publications, a paper published in a book form is sometimes not considered to be a refereed paper. In the case of a journal, the fact that it was given at a symposium may be omitted, and thus, the publication has greater weight in a list of publications. However, it is visible that much fewer papers can be published in a special issue compared to a book and that especially younger scientists are thus hardly represented in special issues. For the "applied mechanician on the street," the publication in book form has the disadvantage that he might not know about the symposium, even if it is in his own field. If he does know about it, he might not have ready access to it. Finally, the book-form publication has a disadvantage for the librarian, who might be unaware of its publication. Even if he is aware of it, he might not be able to fit it into his budget. The proceedings of symposia that are published in a well established journal, even in a special issue, are of no financial consequence to the librarian, for he is paying for a full year's subscription. The trend to move away from books and journals on paper and, instead, to have only electronic publications will further change the situation. Another current issue is, that journals are under high pressure not to publish special issues anymore related to conferences since this might lead to loosing their impact factors or to being banned from major indexing services.

## 3. Length

The maximum and minimum lengths of the symposium proceedings are determined not so much by the symposium organizers but rather by the publishers. Springer normally wanted a 400-page limit. There are exceptions, e.g., where external funds are sponsoring the printing. In such cases, the book publisher has less to say. Typically special issues of journals can only accommodate less than 10 papers and so just a very limited selection of the research presented at a symposia can be included there.

# H. Sponsors and Arrangements

## 1. Organizers

The organizers of the symposia are mostly members of academia, and the institutions involved, other than IUTAM itself, are universities. Rarely there was considerable amount of industrial support of the symposia. There is no symposium in recent memory that was organized by an industrial organization. Research institute sponsorship has sometimes been approved.

## 2. Socials

The duration of the symposia is at least two days and goes up to five days; thus, the inclusion of socials is important. The majority of past symposia have made efforts to accomplish this, as the socials contribute much to the exchange of scientific information.

## 3. Photos of Participants and List of Participants

Some symposia have generated a printed list of participants in a bound volume, yet others have not. A group picture of up to sixty people is quite feasible if the participants are standing on steps in four or five rows. This has been done at certain symposia, and a key identifying the persons was sometimes supplied.

# General Assemblies

Daniel C. Drucker, Stephen Juhasz and Peter Eberhard

## A. When and Where

The General Assembly of the International Union of Theoretical and Applied Mechanics has met every other year since the founding of the Union in 1946. Every second General Assembly has been held in conjunction with the International Congress which is attended by almost all members. The intermediate meetings have been held by invitation at a reasonably convenient and interesting location. Of the 36 meetings listed in the appendix, 7 were held in countries that hosted one each. The remaining 29 took place in 10 other countries.

Present practice is to schedule each meeting for two days. All issues of importance are raised on the first day with final decision taken on the second. This allows ample time for the participants to discuss and become fully familiar with any complex or difficult matters that arise from time to time and to reach consensus prior to the formal vote. One day meetings had been held in the early days. However, when the number of symposia increased dramatically and the General Assembly began to play a much stronger role in the direction of the affairs of the Union, one day did not provide sufficient time for the members either to form firm conclusions in response to recommendations of the Bureau, the Symposia Panels,

D.C. Drucker (1918–2001)
University of Florida, Gainesville, USA

S. Juhasz (1913–2013)
Southwest Research Institute, San Antonio, USA

P. Eberhard (*1966)
University of Stuttgart, Stuttgart, Germany

© The Author(s) 2016
P. Eberhard and S. Juhasz (eds.), *IUTAM*,
DOI 10.1007/978-3-319-31063-3_9

and the Electoral Committee or to propose other matters of interest to them. Meetings have been remarkably harmonious over the years. The General Assembly is a body where lasting friendships as well as fruitful professional interactions develop.

The two day duration has provided an additional side benefit for the intermediate meetings. Periods now are available for stimulating scientific lectures or other activities of professional interest. These make a General Assembly meeting between congresses intellectually rewarding as well as a necessary gathering for the transaction of the essential business of the Union.

## B. Membership

The membership of the General Assembly consists of the appointed representatives of the adhering organizations, representatives of the affiliated organizations, and persons elected by the General Assembly as Member-at-Large. In the early days of IUTAM also personal members were selected, some even for lifetime. Since the Pallanza General Assembly in 1950 there has been a shift toward a membership that is mainly representative. Members-at-Large now are elected for a period of four years from one congress to the next and may be reelected if the members of the General Assembly so choose. The current General Assembly (Report 2014) consists of 123 individuals, including 97 persons representing Adhering Organizations, 14 members without voting right representing Affiliated Organizations as well as 12 elected Members-at-Large. This contrasts with a 32 member General Assembly as listed in the first IUTAM Report. Adhering and Affiliated Organizations are listed in the appendices as well as all General Assembly members of the past and present.

## C. Activities

The agenda for the first day of a General Assembly meeting contains reports for the information of the members, and reports that recommend action to be taken on the second day. Among the important items requiring action at each meeting are the choice of symposia to sponsor from among the many proposed for the upcoming two years hence, and the fixing of the unit dues for that same time period. The election of officers and ordinary members of the Bureau occurs every four years at the time of a congress. Recommendations of the Electoral Committee, based upon nominations solicited six or more months earlier from each member of the General Assembly, are presented on the first day of the meeting.

A typical agenda for a meeting held in conjunction with a congress follows.

## Agenda—21 August, 2012

1. Opening of the meeting by the President
2. Report by the Secretary-General
3. Report by the Treasurer on financial matters
4. Preliminary discussion on annual dues
5. Report by the Secretary of the Congress Committee
6. Matters concerning Adhering Organizations
7. Matters concerning Affiliated and Associate Organizations
8. Reports and preliminary discussions on Working Parties
9. Report of the Electoral Committee
10. Proposals for election of Members-at-Large
11. Proposals for election of members of Symposia Panels
12. Preliminary discussion on future IUTAM Symposia.
13. Preliminary discussion on future Int. Summer Schools on Mechanics
14. Preliminary discussion on a resolution for the eligibility for election as Bureau Officer and as Bureau member
15. Preliminary discussion on a change of statutes concerning the Congress Committee
16. Publication of Proceedings
17. IUTAM Prizes for Fluid Mechanics and for Solid Mechanics.

## Agenda—22 August, 2012

18. Presentation of ICSU by Prof. Dov Jaron, member of ICSU Executive Board
19. Matters concerning Inter-Union Committees
20. Matters concerning non-ICSU Organizations
21. Future IUTAM co-sponsored events
22. Continued discussion and final decision regarding future IUTAM Symposia
23. Continued discussion and final decision regarding future International Summer Schools on Mechanics
24. Continued discussion and final decision regarding annual dues
25. Continued discussion and final decision regarding Working Parties
26. Election of Officers and members of the Bureau
27. Election of Members-at-Large
28. Election of members of the Congress Committee of IUTAM
29. Final decision on re-appointment of four members of Symposia Panels and the election of two new members for the Solid Panel
30. Continued discussion and final decision regarding the resolution for the eligibility for election as Bureau Officer and as Bureau member
31. Date and venue of the next General Assembly
32. Any other business.

The agenda for an intermediate meeting would not have the Electoral Committee report and the elections. Several topics arise just from time to time, such as admission of Adhering and Affiliated organizations, revision of Statutes, report of a study group or other issues.

# Reports and Further References

Stephen Juhasz, Frithiof I. Niordson and Peter Eberhard

## A. Frequency, Availability

The IUTAM Reports were issued annually with one exception. There was a combined three-year report for the years 1954–55–56. IUTAM was formally established in 1946 and the first issue was published for the year 1948. The reports are usually printed during the calendar year following the year to which they refer. The last available report used for this edition of the book is for the 2014 calendar year. Copies of reports have been available from the IUTAM Secretariat and all reports are now available in scanned form on the IUTAM website www.iutam.org.

S. Juhasz (1913–2013)
Southwest Research Institute, San Antonio, USA

F.I. Niordson (1922–2009)
Technical University of Denmark, Lyngby, Denmark

P. Eberhard (*1966)
University of Stuttgart, Stuttgart, Germany

© The Author(s) 2016
P. Eberhard and S. Juhasz (eds.), *IUTAM*,
DOI 10.1007/978-3-319-31063-3_10

# B. Standardization

Unlike the IUTAM Proceedings, efforts have been made from the beginning, most likely by the General Secretaries, to achieve standardization in IUTAM Reports. The first phase was the standardization of size of the report and of the cover (16 × 23.5 mm, with ivory color). The second phase was the standardization of the information on the cover.

1. Information on title page in, English (as is majority of reports)
2. On top "International Union of Theoretical and Applied Mechanics."
3. In the center "Report..."
4. On the bottom the IUTAM logo with the address of the Secretariat.

The third phase of the standarization dealt with the content. Until 1968 the content of the report centered around the Bureau and General assembly meetings, followed by the appendices in somewhat arbitrary fashion. With the 1969 report the grouping and sequencing has been improved. Also, an alphabetical "List of Addresses" of persons associated with IUTAM was introduced. From 1984, the spine of the volume is lettered "IUTAM" and also with "Report 1984."

In the current format the first part is devoted to IUTAM activities with the record of the Annual General Assembly meetings given only in summary form. The second part contains reports with one or two exceptions relating to affiliated and adhering organizations and also detailed reports about all symposia of the corresponding year. The third part contains IUTAM statutes, resolutions, rules and procedures. This section concludes with a list of publications and list of names and addresses. The consecutive reports basically maintained the above format.

# C. Further References

Bibliographical details of Proceedings of IUTAM Congresses, Symposia, and Reports are provided in the appendix. They have been the main sources of information for this book. Further interesting reading related to the history of IUTAM can be found in the following references.

Alkemade, A.J.Q., IUTAM 1946-1996; Fifty Years of Impulse to Mechanics. Dordrecht, Kluwer, 1996.

Battimelli, G., The Mathematician and the Engineer: Statistical Theories of Turbulence in the 20's. Rivista di storia delle scienze mediche e naturali 1:73-94, 1984.

Drucker, D.C. and Lee, E.H., Notes on the Eighth International Congress on Theoretical and Applied Mechanics, August 20 to 28, 1952, Istanbul, Turkey. Applied Mechanics Reviews 5: 497-498, 1952.

Goodstein, J.R., and Kopp, C., eds. The Theodore von Karman Collection at the California Institute of Technology; Guide to the original collection and a microfiche edition. Pasadena, California Institute of Technology, 1981. (also at resolver.caltech.edu/CaltechBOOK:1981.004)

Juhasz, S., Famous Mechanics Scientists; Notes on AMR-RZM Exhibit/XIII IUTAM Congress, Moscow 1972. Applied Mechanics Reviews 26: 145-160, 1973.

Juhasz, S., From Delft to Delft. (AMR Report no. 59). San Antonio, Applied Mechanics Reviews, 1976.

Juhasz, S., International Congresses of Theoretical and Applied Mechanics: The First Half Century. Applied Mechanics Reviews 30: 1323-1324, 1977.

Juhasz, S., Mechanics in Action; XV IUTAM Congress, Toronto 1980. Applied Mechanics Reviews 34: 317, 1981.

Schiehlen, W. and van Wijngaarden, L., eds. Mechanics at the Turn of the Century. Aachen, Shaker Verlag, 2000.

von Karman, T. and Edson, L., The Wind and Beyond, pp. 104-105 and pp. 294-295. Boston, Little, Brown and Co., 1967.

# Appendix 1
# Congresses by Number, Location, and Year

| 0 | Innsbruck 1922 |
|---|---|

| | | | | |
|---|---|---|---|---|
| I | Delft 1924 | XIII | Moscow 1972 |
| II | Zurich 1926 | XIV | Delft 1976 |
| III | Stockholm 1930 | XV | Toronto 1980 |
| IV | Cambridge UK 1934 | XVI | Lyngby 1984 |
| V | Cambridge USA 1938 | XVII | Grenoble 1988 |
| VI | Paris 1946 | XVIII | Haifa 1992 |
| VII | London 1948 | XIX | Kyoto 1996 |
| VIII | Istanbul 1952 | XX | Chicago 2000 |
| IX | Bruxelles 1956 | XXI | Warsaw 2004 |
| X | Stresa 1960 | XXII | Adelaide 2008 |
| XI | Munich 1964 | XXIII | Beijing 2012 |
| XII | Stanford 1968 | XXIV | Montreal 2016 |

map: www.commons.wikimedia.org

© The Author(s) 2016
P. Eberhard and S. Juhasz (eds.), *IUTAM*,
DOI 10.1007/978-3-319-31063-3

# Appendix 2
# Congress Participants by Countries

0 = Country was not represented    - = Country existed not yet or no longer

| | 0 1922 | I 1924 | II 1926 | III 1930 | IV 1934 | V 1938 | VI 1946 | VII 1948 | VIII 1952 |
|---|---|---|---|---|---|---|---|---|---|
| Argentina | 0 | 0 | 0 | 0 | 0 | 2 | 0 | 1 | 0 |
| Australia | 0 | 1 | 0 | 0 | 0 | 0 | 4 | 8 | 2 |
| Austria | 3 | 2 | 6 | 14 | 1 | 3 | 0 | 1 | 1 |
| Belgium | 0 | 2 | 3 | 6 | 4 | 0 | 19 | 11 | 5 |
| Brazil | 0 | 0 | 0 | 1 | 0 | 0 | 1 | 0 | 0 |
| Bulgaria | 0 | 1 | 1 | 3 | 1 | 1 | 2 | 1 | 0 |
| Canada | 0 | 1 | 0 | 2 | 1 | 7 | 7 | 1 | 0 |
| China | 0 | 0 | 0 | 3 | 0 | 0 | 2 | 0 | 0 |
| CSSR | 2 | 4 | 8 | 12 | 2 | 0 | 13 | 19 | 0 |
| Denmark | 0 | 0 | 0 | 16 | 2 | 0 | 2 | 7 | 4 |
| Egypt | 0 | 1 | 2 | 2 | 2 | 1 | 1 | 6 | 4 |
| Finland | 0 | 0 | 0 | 5 | 0 | 0 | 0 | 1 | 2 |
| France | 0 | 3 | 14 | 39 | 26 | 14 | 108 | 60 | 35 |
| Germany | 14 | 49 | 66 | 135 | 37 | 22 | 0 | 0 | 25 |
| Greece | 0 | 0 | 1 | 3 | 0 | 0 | 0 | 0 | 3 |
| Hungary | 0 | 1 | 0 | 4 | 3 | 0 | 0 | 1 | 0 |
| India | 0 | 0 | 0 | 0 | 1 | 0 | 1 | 3 | 2 |
| Iran | 0 | 0 | 0 | 0 | 0 | 0 | 0 | 2 | 1 |
| Ireland | 0 | 0 | 0 | 2 | 0 | 0 | 1 | 2 | 0 |
| Israel | 0 | 0 | 0 | 0 | 0 | 0 | 0 | 0 | 2 |
| Italy | 5 | 3 | 5 | 10 | 2 | 1 | 19 | 26 | 12 |
| Japan | 0 | 0 | 4 | 24 | 6 | 5 | 0 | 0 | 3 |
| Netherlands | 6 | 103 | 10 | 17 | 9 | 2 | 24 | 48 | 16 |
| Norway | 1 | 3 | 1 | 10 | 0 | 0 | 0 | 4 | 4 |
| Poland | 0 | 3 | 6 | 10 | 5 | 3 | 4 | 14 | 0 |
| Portugal | 0 | 0 | 0 | 0 | 0 | 0 | 1 | 0 | 0 |
| Romania | 0 | 2 | 3 | 10 | 1 | 0 | 13 | 0 | 0 |
| South Africa | 0 | 0 | 0 | 0 | 0 | 0 | 0 | 1 | 0 |
| Spain | 0 | 0 | 1 | 1 | 1 | 0 | 2 | 1 | 1 |
| Sweden | 2 | 2 | 5 | 113 | 6 | 2 | 11 | 16 | 12 |
| Switzerland | 0 | 1 | 91 | 9 | 6 | 1 | 18 | 19 | 4 |
| Turkey | 0 | 1 | 2 | 3 | 2 | 3 | 6 | 7 | 77 |
| U. Kingdom | 0 | 19 | 13 | 37 | 114 | 34 | 141 | 491 | 24 |
| USA | 0 | 1 | 4 | 51 | 18 | 287 | 114 | 87 | 90 |
| USSR | 0 | 3 | 0 | 21 | 3 | 3 | 1 | 0 | 0 |
| Yugoslavia | 0 | 1 | 1 | 3 | 0 | 0 | 0 | 8 | 3 |
| others | 0 | 0 | 0 | 17 | 2 | 4 | 0 | 2 | 2 |
| total | 33 | 207 | 247 | 583 | 255 | 395 | 515 | 848 | 334 |

© The Author(s) 2016
P. Eberhard and S. Juhasz (eds.), *IUTAM*,
DOI 10.1007/978-3-319-31063-3

| | IX 1956 | X 1960 | XI 1964 | XII 1968 | XIII 1972 | XIV 1976 | XV 1980 | XVI 1984 |
|---|---|---|---|---|---|---|---|---|
| Algeria | 3 | 1 | 0 | 0 | 0 | 0 | 3 | 1 |
| Argentina | 0 | 0 | 1 | 0 | 1 | 0 | 0 | 0 |
| Australia | 0 | 1 | 2 | 13 | 19 | 6 | 3 | 6 |
| Austria | 2 | 2 | 14 | 5 | 4 | 8 | 4 | 4 |
| Belgium | 33 | 15 | 13 | 8 | 8 | 18 | 9 | 11 |
| Brazil | 1 | 0 | 0 | 2 | 0 | 2 | 2 | 3 |
| Bulgaria | 1 | 6 | 2 | 0 | 14 | 6 | 3 | 2 |
| Canada | 3 | 2 | 16 | 45 | 28 | 24 | 111 | 32 |
| China | 5 | 1 | 0 | 0 | 0 | 0 | 12 | 37 |
| CSSR | 15 | 9 | 9 | 4 | 69 | 18 | 3 | 8 |
| Denmark | 3 | 11 | 13 | 7 | 13 | 19 | 10 | 75 |
| Egypt | 0 | 0 | 1 | 0 | 0 | 1 | 2 | 1 |
| Finland | 2 | 2 | 2 | 0 | 2 | 2 | 1 | 5 |
| France | 80 | 100 | 80 | 61 | 63 | 115 | 54 | 63 |
| Germany/FRG | 53 | 89 | 330 | 50 | 52 | 84 | 39 | 43 |
| Germany/GDR | - | - | - | - | 25 | 5 | 3 | 3 |
| Greece | 2 | 2 | 3 | 0 | 1 | 4 | 1 | 4 |
| Hungary | 1 | 2 | 7 | 1 | 15 | 6 | 1 | 3 |
| India | 2 | 2 | 3 | 4 | 24 | 15 | 7 | 3 |
| Iran | 1 | 0 | 0 | 1 | 2 | 2 | 0 | 3 |
| Ireland | 1 | 0 | 2 | 0 | 4 | 3 | 0 | 2 |
| Israel | 3 | 7 | 4 | 4 | 0 | 16 | 19 | 28 |
| Italy | 18 | 55 | 15 | 10 | 22 | 23 | 16 | 26 |
| Japan | 16 | 9 | 11 | 21 | 29 | 32 | 33 | 43 |
| Korea | 0 | 2 | 0 | 1 | 0 | 0 | 1 | 0 |
| Mexico | 0 | 0 | 0 | 0 | 0 | 1 | 2 | 2 |
| Netherlands | 47 | 43 | 46 | 36 | 30 | 212 | 34 | 39 |
| New Zealand | 0 | 0 | 0 | 1 | 1 | 0 | 2 | 1 |
| Nigeria | 0 | 0 | 0 | 1 | 0 | 2 | 3 | 1 |
| Norway | 4 | 4 | 5 | 2 | 2 | 5 | 0 | 4 |
| Poland | 26 | 16 | 9 | 15 | 64 | 12 | 11 | 17 |
| Portugal | 0 | 1 | 0 | 0 | 1 | 4 | 3 | 1 |
| Romania | 9 | 5 | 5 | 8 | 14 | 3 | 0 | 4 |
| Saudi Arabia | 0 | 0 | 0 | 0 | 0 | 1 | 2 | 2 |
| South Africa | 1 | 0 | 0 | 0 | 0 | 2 | 1 | 1 |
| Spain | 2 | 1 | 2 | 0 | 0 | 1 | 1 | 3 |
| Sweden | 15 | 21 | 12 | 20 | 26 | 39 | 21 | 38 |
| Switzerland | 10 | 9 | 12 | 3 | 2 | 16 | 7 | 6 |
| Turkey | 13 | 2 | 6 | 1 | 8 | 8 | 2 | 3 |
| U. Kingdom | 137 | 116 | 156 | 73 | 86 | 79 | 47 | 54 |
| USA | 137 | 196 | 133 | 879 | 177 | 159 | 247 | 183 |
| USSR | 14 | 51 | 30 | 48 | 1411 | 35 | 9 | 11 |
| Venezuela | 0 | 0 | 1 | 4 | 2 | 0 | 1 | 0 |
| Yugoslavia | 15 | 14 | 11 | 2 | 26 | 13 | 4 | 2 |
| others | 0 | 0 | 4 | 2 | 5 | 4 | 4 | 2 |
| total | 675 | 797 | 960 | 1332 | 2250 | 1005 | 738 | 780 |

|  | XVII 1988 | XVIII 1992 | XIX 1996 | XX 2000 | XXI 2004 | XXII 2008 | XXIII 2012 |
|---|---|---|---|---|---|---|---|
| Armenia | - | 0 | 0 | 3 | 1 | 1 | 1 |
| Australia | 4 | 4 | 11 | 9 | 10 | 139 | 22 |
| Austria | 5 | 3 | 10 | 23 | 18 | 8 | 8 |
| Azerbaijan | - | 1 | 0 | 0 | 0 | 0 | 0 |
| Belarus | - | 0 | 2 | 0 | 4 | 1 | 4 |
| Belgium | 6 | 5 | 8 | 6 | 19 | 6 | 5 |
| Brazil | 7 | 2 | 2 | 6 | 13 | 6 | 10 |
| Bulgaria | 8 | 5 | 2 | 3 | 7 | 0 | 1 |
| Cameroon | 1 | 0 | 0 | 0 | 0 | 0 | 0 |
| Canada | 25 | 18 | 16 | 23 | 19 | 14 | 21 |
| Chile | 1 | 0 | 0 | 1 | 2 | 0 | 2 |
| China | 28 | 8 | 19 | 29 | 54 | 120 | 400 |
| China, Taipei | 8 | 6 | 19 | 6 | 18 | 30 | 14 |
| Colombia | 0 | 0 | 0 | 0 | 0 | 0 | 1 |
| Croatia | - | - | 0 | 0 | 1 | 1 | 1 |
| CSSR | 6 | 3 | - | - | - | - | - |
| Czech Republic | - | - | 4 | 4 | 17 | 9 | 2 |
| Cuba | 0 | 0 | 0 | 1 | 0 | 0 | 0 |
| Cyprus | 0 | 0 | 0 | 0 | 0 | 1 | 2 |
| Denmark | 10 | 9 | 25 | 28 | 30 | 26 | 26 |
| Egypt | 0 | 0 | 2 | 0 | 0 | 1 | 0 |
| Estonia | - | 0 | 2 | 2 | 8 | 4 | 4 |
| Finland | 11 | 1 | 4 | 4 | 12 | 6 | 1 |
| France | 340 | 39 | 50 | 116 | 170 | 89 | 89 |
| Georgia | - | 0 | 0 | 1 | 0 | 2 | 2 |
| Germany/FRG | 45 | 36 | 50 | 101 | 118 | 69 | 64 |
| Germany/GDR | 2 | - | - | - | - | - | - |
| Greece | 5 | 4 | 2 | 5 | 9 | 5 | 6 |
| Guade Loupe | 0 | 0 | 0 | 1 | 0 | 0 | 0 |
| Hungary | 5 | 2 | 4 | 3 | 10 | 4 | 4 |
| India | 3 | 0 | 4 | 8 | 12 | 25 | 19 |
| Iran | 0 | 0 | 0 | 0 | 8 | 4 | 2 |
| Ireland | 1 | 1 | 1 | 5 | 7 | 2 | 6 |
| Israel | 23 | 85 | 19 | 33 | 37 | 10 | 14 |
| Italy | 15 | 11 | 17 | 36 | 25 | 18 | 23 |
| Japan | 33 | 22 | 332 | 94 | 89 | 84 | 76 |
| Jordan | 0 | 0 | 0 | 0 | 0 | 1 | 0 |
| Kazakhstan | - | 0 | 0 | 0 | 2 | 2 | 5 |
| Korea | 0 | 3 | 11 | 9 | 19 | 19 | 8 |
| Kuwait | 1 | 0 | 0 | 0 | 0 | 0 | 0 |

Appendix 2: Congress Participants by Countries

| | XVII 1988 | XVIII 1992 | XIX 1996 | XX 2000 | XXI 2004 | XXII 2008 | XXIII 2012 |
|---|---|---|---|---|---|---|---|
| Kyrgyzstan | - | 0 | 0 | 0 | 0 | 1 | 0 |
| Latvia | - | 0 | 2 | 2 | 7 | 2 | 3 |
| Lebanon | 0 | 0 | 0 | 0 | 0 | 1 | 0 |
| Lithuania | - | 0 | 1 | 0 | 3 | 0 | 0 |
| Macau | 0 | 0 | 0 | 0 | 0 | 1 | 1 |
| Malaysia | 0 | 0 | 0 | 0 | 0 | 4 | 1 |
| Mexico | 2 | 1 | 1 | 1 | 5 | 1 | 5 |
| Moldavia | - | 0 | 0 | 0 | 0 | 1 | 0 |
| Nepal | 0 | 0 | 0 | 1 | 0 | 0 | 0 |
| Netherlands | 34 | 21 | 33 | 28 | 43 | 25 | 21 |
| New Zealand | 3 | 1 | 3 | 1 | 1 | 5 | 3 |
| Nigeria | 1 | 0 | 0 | 0 | 1 | 1 | 0 |
| Norway | 3 | 1 | 2 | 3 | 8 | 2 | 7 |
| Poland | 23 | 21 | 21 | 25 | 194 | 33 | 19 |
| Portugal | 5 | 5 | 7 | 3 | 9 | 5 | 4 |
| Romania | 1 | 1 | 0 | 2 | 5 | 3 | 3 |
| Russia | - | 32 | 27 | 48 | 79 | 40 | 64 |
| Saudi Arabia | 1 | 0 | 0 | 2 | 2 | 0 | 2 |
| Serbia | - | - | 2 | 1 | 3 | 2 | 2 |
| Singapore | 0 | 0 | 1 | 8 | 4 | 4 | 10 |
| Slovakia | - | - | 1 | 2 | 1 | 0 | 0 |
| Slovenia | - | - | 1 | 2 | 3 | 2 | 0 |
| South Africa | 0 | 2 | 4 | 7 | 7 | 9 | 5 |
| Spain | 1 | 0 | 0 | 5 | 14 | 6 | 15 |
| Sweden | 44 | 7 | 16 | 23 | 18 | 13 | 12 |
| Switzerland | 14 | 10 | 13 | 24 | 12 | 11 | 9 |
| Thailand | 0 | 0 | 2 | 0 | 0 | 1 | 0 |
| Trinidad | 0 | 1 | 0 | 0 | 0 | 0 | 1 |
| Tunisia | 1 | 0 | 0 | 0 | 0 | 0 | 0 |
| Turkey | 7 | 1 | 0 | 1 | 4 | 1 | 2 |
| UK | 60 | 35 | 65 | 83 | 107 | 62 | 51 |
| Ukraine | - | 4 | 4 | 5 | 38 | 6 | 10 |
| USA | 123 | 107 | 107 | 587 | 204 | 151 | 153 |
| USSR | 29 | - | - | - | - | - | - |
| Uzbekistan | - | 0 | 0 | 0 | 0 | 0 | 0 |
| Venezuela | 1 | 0 | 0 | 0 | 0 | 0 | 0 |
| Vietnam | 0 | 1 | 0 | 0 | 0 | 1 | 0 |
| Yugoslavia | 5 | 5 | - | - | - | - | - |
| total | 951 | 525 | 936 | 1430 | 1515 | 1105 | 1271 |

# Appendix 3
# Congress Participants from "Other" Countries

List of countries not included in the congress participant matrix where delegates participated in few Congresses. The Roman numeral refers to the Congress and the following Arabic number to the number of delegates.

| | |
|---|---|
| Algeria | XXI-1, XXII-1, XXIII-2 |
| Arab Emirates | XXIII-1 |
| Armenia | III-1, IV-1 |
| Chile | V-1 |
| Colombia | VI-1 |
| Congo | XI-1 |
| Cuba | XIII-1 |
| Estonia | III-5, IV-1 |
| Ghana | XIV-1 |
| Hong Kong | XIV-1 |
| Iccland | XIV-1 |
| Indonesia | VII-1 |
| Iraq | XV-1 |
| Latvia | III-7 |
| Lebanon | V-1, VI-1, XI-1 |
| Lithuania | III-3 |
| Libya | XXI-2, XXII-2 |
| Malta | VIII-1 |
| Monaco | XI-2 |
| Pakistan | VIII-1 |
| Palestine | III-2, V-1, VI-2 |
| Peru | XII-1 |
| Qatar | XXI-1, XXII-1, XXIII-1 |
| Senegal | XII-1 |
| Thailand | V-1 |
| Singapore | VII-1, XVI-1 |
| Sri Lanka | XIII-1 |
| Sudan | XIV-1 |
| Tunisia | XV-2, XVI-1 |
| Uruguay | III-1 |
| Vietnam | XIII-3 |
| West Indies | XV-1 |

© The Author(s) 2016
P. Eberhard and S. Juhasz (eds.), *IUTAM*,
DOI 10.1007/978-3-319-31063-3

# Appendix 4
# Congress Participants by Continents

|  | O 1922 | I 1924 | II 1926 | III 1930 | IV 1934 | V 1938 | VI 1946 | VII 1948 | VIII 1952 |
|---|---|---|---|---|---|---|---|---|---|
| Africa | 0 | 1 | 2 | 2 | 2 | 1 | 1 | 7 | 4 |
| Asia | 0 | 1 | 4 | 29 | 7 | 7 | 4 | 5 | 8 |
| Australia | 0 | 1 | 0 | 0 | 0 | 0 | 4 | 8 | 2 |
| Europe | 33 | 202 | 237 | 507 | 227 | 91 | 383 | 739 | 230 |
| North America | 0 | 2 | 4 | 53 | 19 | 293 | 121 | 88 | 90 |
| South America | 0 | 0 | 0 | 2 | 0 | 3 | 2 | 1 | 0 |
| total | 33 | 207 | 247 | 593 | 255 | 395 | 515 | 848 | 334 |

|  | IX 1956 | X 1960 | XI 1964 | XII 1968 | XIII 1972 | XIV 1976 | XV 1980 | XVI 1984 |
|---|---|---|---|---|---|---|---|---|
| Africa | 7 | 0 | 2 | 2 | 0 | 7 | 8 | 5 |
| Asia | 24 | 14 | 15 | 27 | 59 | 51 | 56 | 91 |
| Australia | 0 | 1 | 2 | 13 | 19 | 6 | 3 | 3 |
| Europe | 503 | 584 | 790 | 359 | 1965 | 756 | 307 | 458 |
| North America | 140 | 198 | 149 | 924 | 203 | 183 | 360 | 217 |
| South America | 1 | 0 | 2 | 7 | 4 | 2 | 4 | 3 |
| total | 675 | 797 | 960 | 1332 | 2250 | 1005 | 738 | 777 |

|  | XVII 1988 | XVIII 1992 | XIX 1996 | XX 2000 | XXI 2004 | XXII 2008 | XXIII 2012 |
|---|---|---|---|---|---|---|---|
| Africa | 2 | 3 | 6 | 7 | 4 | 11 | 7 |
| Asia | 105 | 130 | 414 | 199 | 247 | 310 | 573 |
| Australia | 7 | 5 | 14 | 10 | 11 | 144 | 25 |
| Europe | 678 | 259 | 376 | 594 | 1010 | 466 | 473 |
| North America | 148 | 125 | 124 | 612 | 228 | 165 | 174 |
| South America | 11 | 3 | 2 | 8 | 15 | 7 | 19 |
| total | 951 | 525 | 936 | 1430 | 1515 | 1103 | 1271 |

Countries where continental assignment is a question of definition

| USSR / Russia | Europe |
|---|---|
| Iceland | Europe |
| Turkey | Europe |

© The Author(s) 2016
P. Eberhard and S. Juhasz (eds.), *IUTAM*,
DOI 10.1007/978-3-319-31063-3

# Appendix 5
# Congress Participation by Host Country Participants

| | Congress | Host Country Participants | Out of Country Participants | Total | Locals % |
|---|---|---|---|---|---|
| 0 | Innsbruck 1924 | 3 | 30 | 33 | 9 |
| I | Delft 1924 | 103 | 104 | 207 | 50 |
| II | Zurich 1926 | 91 | 156 | 247 | 37 |
| III | Stockholm 1930 | 114 | 479 | 593 | 19 |
| IV | Cambridge UK 1934 | 114 | 141 | 255 | 45 |
| V | Cambridge USA 1938 | 286 | 109 | 395 | 72 |
| VI | Paris 1946 | 108 | 407 | 515 | 21 |
| VII | London 1948 | 491 | 357 | 848 | 58 |
| VIII | Istanbul 1952 | 77 | 257 | 334 | 23 |
| IX | Bruxelles 1956 | 33 | 642 | 675 | 5 |
| X | Stresa 1960 | 55 | 742 | 797 | 7 |
| XI | Munich 1964 | 330 | 630 | 960 | 34 |
| XII | Stanford 1968 | 879 | 453 | 1332 | 66 |
| XIII | Moscow 1972 | 1411 | 839 | 2250 | 63 |
| XIV | Delft 1976 | 212 | 793 | 1005 | 21 |
| XV | Toronto 1980 | 111 | 627 | 738 | 15 |
| XVI | Lyngby 1984 | 75 | 705 | 780 | 10 |
| XVII | Grenoble 1988 | 340 | 611 | 951 | 36 |
| XVIII | Haifa 1992 | 85 | 440 | 525 | 16 |
| XIX | Kyoto 1996 | 332 | 604 | 936 | 35 |
| XX | Chicago 2000 | 587 | 843 | 1430 | 41 |
| XXI | Warsaw 2004 | 194 | 1321 | 1515 | 13 |
| XXII | Adelaide 2008 | 139 | 966 | 1105 | 13 |
| XXIII | Beijing 2012 | 400 | 871 | 1271 | 31 |

© The Author(s) 2016
P. Eberhard and S. Juhasz (eds.), *IUTAM*,
DOI 10.1007/978-3-319-31063-3

# Appendix 6
# General and Sectional Lecturers: Delft 1924

From Proceedings of the First International Congress for Applied Mechanics Delft, 1924, edited by Prof. C.B. Biezeno and Prof. J.M. Burgers

## Part I: Lectures Delivered at the General Meetings

C.B. Biezeno, Graphical and Numerical Methods for Solving Stress Problems

E.G. Coker, Some Engineering Problems of Stress Distribution

A.A. Griffitth, The Use of Soap Films in Solving Stress Problems

L. Prandtl, Spannungsverteilung in plastischen Körpern

A.A. Griffitth, The Theory of Rupture

A.F. Joffe, Plastizität und Festigkeit der Kristalle

J. Czochralski, Die Beziehungen der Metallographie zur physikalischen Forschung

G.I. Taylor, Experiments with Rotating Fluids

T. von Karman, Über die Stabilität der Laminarströmung und die Theorie der Turbulenz

J.M. Burgers, The Motion of a Fluid in the Boundary Layer along a Plane Smooth Surface

T. Levi-Civita, La détermination rigoureuse des ondes permanentes d'ampleur finie

E. Hogner, Über die Theorie der von einem Schiff erzeugten Wellen und des Wellenwiderstandes

Sir Napier Shaw, The Physical Structure of the Atmosphere regarded from the Dynamical Point of View

© The Author(s) 2016
P. Eberhard and S. Juhasz (eds.), *IUTAM*,
DOI 10.1007/978-3-319-31063-3

# Part II: Lectures Delivered at the Sectional Meetings

## Section I. Rational Mechanics

H. Alt, Kinematische Synthese

R. von Mises, Motorrechnung, ein neues Hilfsmittel der Mechanik

J. Droste, Eine Bemerkung zu den Variationsprinzipien der Mechanik und der Physik

I. Tzénoff, Une forme nouvelle des équations du mouvement des systèmes non-holonomes et son application dans la théorie des percussions

P. Frank, Die geometrische Deutung von Painlevé's Theorie der reellen Bahnkurven allgemeiner mechanischer Systeme

A.N. Kriloff, On the numerical integration of differential equations

H. Föttinger, Über Maschinen zur Integration von Wirbel- und Quellfunktionen (Vektor-Integratoren)

R. Courant, Über direkte Methoden der Variationsrechnung zur Lösung von Randwertaufgaben

H.P. Berlage Jr., Über die Berechnung der Bodenbewegung aus dem Seismogram

F.A. Vening Meinesz, Schwerkraftsbestimmung auf dem Ozean mittels Pendelbeobachtungen in einem Unterseeboote

F. Pfeiffer, Sperrungsvorgänge bei Gleitreibung starrer Körper

## Section II. Theory of Elasticity

E. Schwerin, Die Torsionsstabilität des dünnwandigen Rohres

R.V. Southwell, Note on the Stability under Shearing Forces of a Flat Elastic Strip, and an Analogy with the Problem of Laminar Fluid Motion

R. Grammel, Die Knickung von Schraubenfedern

W. Hort, Analytische Ermittlung der Eigentöne verjüngter Stäbe

K. Terzaghi, Die Theorie der hydrodynamischen Spannungserscheinungen und ihr erdbautechnisches Anwendungsgebiet

H. Reissner, Zum Erddruckproblem

H. Hencky, Zur Theory plastischer Deformationen und der hierdurch im Material hervorgerufenen Nebenspannungen

A. Nádai, Beobachtungen der Gleitflächenbildung an plastischen Stoffen

B.P. Haigh, Theory of Rupture in Fatigue

G. Masing, Das Aufreissen von Messing durch innere Spannungen

E. Schmid, Neuere Untersuchungen an Metallkristallen

T. Wyss, Experimentelle Spannungsuntersuchungen an einem hakenförmigen Körper

J. Geiger, Messgeräte und Verfahren zur Untersuchung mechanischer Schwingungsvorgänge

## Section III. Hydro- and Aerodynamics

N. Zeilon, On Potential Problems in the Theory of Fluid Resistance

U. Cisotti, Sur les mouvements de rotation d'un liquide visqueux

S. Brodetsky, Vortex Motion

T. Rehbock, Die Wasserwalzen als Regler des Energie-Haushaltes der Wasserläufe

H. Solberg, Zum Turbulenzproblem

L. Keller und A. Friedmann, Differentialgleichungen für die turbulente Bewegung einer kompressiblen Flüssigkeit

N. Kotschin, Über starke Diskontinuitäten in einer kompressiblen Flüssigkeit

V. Bjerknes, Die hydrodynamischen Fernkräfte und deren Zusammenhang mit den Auftriebskräften von welchen die Aeroplane getragen werden

C. Koning, Einige Bemerkungen über nichtstationäre Strömungen an Tragflügeln

C. Witoszynski, Modification du principe de circulation

E. Hahn, Note sur l'application aux turbomachines des théories moderne de l'hydrodynamique

M.M. Munk, The Simplifying Assumptions, Reducing the Strict Application of Classical Hydrodynamics to Practical Aeronautical Computation

G. Kempf, Über den Reibungswiderstand von Flächen verschiedener Form

G.V. Baumhauer, Some Notes on Helicopters

# Appendix 7
# General and Keynote Lecturers
# with Lecture Titles

It should be noted that for the Congresses VII–XVI due to missing data the sectional lecturers are not included in this table while for the other Congresses I–VI and XVII–XXIII they are

| Name | Title | Congress |
|------|-------|----------|
| Alfvén, H. | Space research and the new approach to the mechanics of fluid media in cosmos | XVI |
| Ambrósio, J. | Multibody dynamics: Bridging for multidisciplinary applications | XXI |
| Argyris, J.H. | Computer and mechanics | XIV |
| Arnold, V.I. | Bifurcations and singularities in mathematics and mechanics | XVII |
| Bairstow, L. | Airscrew theory - a summary | III |
| Bajer, K. | Rapid formation of strong gradients and diffusion in the transport of scalar and vector fields | XXI |
| Barenblatt, G.I. | Micromechanics of fracture | XVIII |
| Batchelor, G.K. | Recent developments in turbulence research | VII |
| Batchelor, G.K. | Developments in microhydrodynamics | XIV |
| Bau, H.H. | Controlling chaotic convection | XVIII |
| Benjamin, T.B. | Fluid flow with flexible boundaries | XI |
| Beysens, D.A. | Near-critical point hydrodynamics and microgravity | XXI |
| Bhattacharya, K. | Defects in crystalline solids: Manifestation of quantum mechanics at continuum scales | XXIII |
| Biezeno, C.B. | Graphical and numerical methods for solving stress problems | I |
| Bigoni, D. | Material instabilities in elastic and plastic solids: The perturbative approach | XXII |
| Bogolioubov, N. | Methodes analytiques de la theorie des oscillations non-linéaires | X |
| Brady, J.F. | Suspensions: from micromechanics to macroscopic behavior | XXI |
| Brennert, M. | Droplet splashing | XXIII |
| Bridgman, P.W. | Some mechanical properties of matter under high pressure | II |

P. Eberhard and S. Juhasz (eds.), *IUTAM*,
DOI 10.1007/978-3-319-31063-3

| Name | Title | Congress |
|---|---|---|
| Broer, L.J.F. | Some recent advances in the theory of wave propagation | XI |
| Bryson, A.E. | Control theory for random systems | XIII |
| Burgers, J.M. | The motion of a fluid in the boundary layer along a plane smooth surface | I |
| Burzio, F. | Soluzioni sperimentali del secondo problema balistico | IV |
| Bush, V. | Recent progress in analysing machines | IV |
| Calladine, C.R. | Application of structural mechanics to biological systems | XVIII |
| Camichel, C. | Sur la théorie des coups de belier | II |
| Carrier, G.F. | Phenomena in rotating fluids | XI |
| Casimir, H.B.G. | Mathematics, mechanics and our conception of the physical world | XIV |
| Chatfield, C.H. | Applied mechanics in aeronautical engineering | V |
| Chen, S. | Multiscale fluid mechanics and modeling | XXIII |
| Cheng, G.-D. | Some development of structural topology optimization | XIX |
| Coker, E.G. | Some engineering problems of stress distribution | I |
| Colonnetti, G. | Non-linear deformations of solid bodies | X |
| Corigliano, A. | Microsystems and mechanics | XXIII |
| Couder, Y. | Viscous fingering as a pattern forming system | XVIII |
| Courant, R. | Reciprocal variational problems and applications to problems of equilibrium | VIII |
| Cowley, S.J. | Laminar boundary-layer theory: A 20$^{th}$ century paradox? | XX |
| Cross, H. | The relation of structural mechanics to structural engineering | V |
| Czochralski, J. | Die Beziehungen der Metallographie zur physikalischen Forschung | I |
| d'Adhemar, R. | Étude du mouvement pendulaire d'un projectile toumant | III |
| Dagan, G. | Effective, equivalent and apparent properties of heterogeneous media | XX |
| Davidson, K.S.M. | Ships | IX |

| Name | Title | Congress |
|------|-------|----------|
| de Borst, R. | Multi-scale mechanics and evolving discontinuities: computational issues | XXIII |
| Debye, P. | Molekulare Kräfte und ihre Deutung | II |
| Durrant-Whyte, H. | Maximal information systems | XXII |
| Eberhard, P. | Particles - bridging the gap between solids and fluids | XXIII |
| Eisner, F. | Das Widerstandsproblem | III |
| Elad, D. | Biomechanical aspects in human reproduction | XXII |
| Ferri, A. | Some heat transfer problems near stagnation region of blunt bodies at hypersonic speed | X |
| Fichera, G. | Analytic problems of materials with memory | XV |
| Fleck, N. | Micro-architectured solids: from blast resistant structures to morphing wings | XXII |
| Fraeijs deVeubeke, B. | The numerical analysis of structures | XIII |
| Freund, L.B. | Entropic forces in the mechanics of solids | XXIII |
| Fung, Y.C. | Biomechanics | XIV |
| Gao, H. | Nanoscale mechanics of biological materials | XXI |
| Gauthier, L. | Sur le flambement et les questions d'instabilité | XII |
| Germain, P. | Quelques progres recents en aerodynamique théorique des grandes vitesses | IX |
| Gharib, M. | Lessons for bio-inspired engineering: fluid mechanics of embryonic heart | XXIII |
| Goldstein, R. | Scale interaction and ordering effects at fracture | XXIII |
| Gollub, J.P. | Bifurcations and modal interactions in fluid mechanics: surface waves | XVII |
| Griffith, A.A. | The theory of rupture | I |
| Griffith, A.A. | The use of soap films in solving stress problems | I |
| Grimvall, G. | Mechanics in sport | XVIII |
| Gupta, N.K. | Plasto-mechanics of large deformation under impact loading | XXII |
| Hagedorn, P. | Active vibration damping in large flexible structures | XVII |

| Name | Title | Congress |
|---|---|---|
| Herring, J.R. | Numerical simulation of turbulence | XVII |
| Hetenyi, M. | The present state of development of experimental stress analysis | VII |
| Hilgenfeldt, S. | Cellular matter: interfacial mechanics and geometry | XXII |
| Hogner, E. | Über die Theorie der von einem Schiff erzeugten Wellen und des Wellenwiderstandes | I |
| Hopfinger, E.J. | Turbulence and vortices in rotating fluids | XVII |
| Hu, W.R. | Onset of oscillatory thermocapillary convention | XXII |
| Hughes, T.J.R. | Variational and multiscale methods in turbulence | XXI |
| Hunsaker, J.C. | Social aspect of aeronautics | VIII |
| Hunt, J.C.R. | Environmental fluid mechanics | XV |
| Hussain, F. | Nonlinear transient growth on a vortex column | XXII |
| Hutchinson, J. | The role of mechanics in advancing thermal barrier coatings | XXII |
| Hutchinson, J. | Mechanisms of toughening in ceramics | XVII |
| Imberger, J. | Physical limnology: advances and future challenges | XXII |
| Iyengar, R.N. | Probabilistic methods in earthquake engineering | XX |
| Jensen, O.E. | Instabilities of flows through deformable tubes and channels | XXII |
| Joffe, A.F. | Plastizität und Festigkeit der Kristalle | I |
| Jones, B.M. | The control of stalled aeroplanes | II |
| Jouguet, E. | La théorie thermodynamique de la propagation des explosions | II |
| Kambe, T. | Aerodynamic sound associated with vortex motions: observation and computation | XVIII |
| Kampé de Fériet, J. | Le tenseur spectral de la turbulence homogène non isotrope dans un fluide incompressible | VII |
| Kantrowitz, A. | Physical phenomena associated with strong shock waves | VIII |

| Name | Title | Congress |
|------|-------|----------|
| Keller, J.B. | Computers and chaos in mechanics | XVI |
| Keunings, R. | Non-Newtonian fluid mechanics using molecular theory | XXI |
| Kida, S. | Vortical structure in turbulence | XX |
| Kieffer, S.W. | Multiphase flow in explosive volcanic and geothermal eruptions | XVII |
| Kiya, M. | Separation bubbles | XVII |
| Kobori, T. | Structural control for large earthquakes | XIX |
| Koerber, F. | Das Verhalten metallischer Werkstoffe im Bereich kleiner Verformungen | V |
| Korn, A. | Automatische Herstellung der Jacquard-Karten für die mechanische Weberei | III |
| Krajcinovic, D. | Essential structure of damage mechanics theorics | XIX |
| Kreuzer, E. | Nonlinear dynamics in ocean engineering | XXI |
| Ladevéze, P.J. | A bridge between the micro- and mesomechanics of laminates, fantasy or reality | XXI |
| Lavrentiev, M.A. | Les problèmes de l'hydrodynamique et les modèles mathématiques | XII |
| Legendre, R. | Ecoulements instationnaires autour d'obstacles en vibration | XI |
| Leipholz, H.H.E. | Analysis of non-conservative, nonholonomic systems | XV |
| Lesieur, M. | Turbulence and large-eddy simulations | XXI |
| Levi-Civita, T. | La détermination rigoureuse des ondes permanents d'ampleur finie | I |
| Levi-Civita, T. | Sur les chocs dans le problème des trois corps | II |
| Libai, A. | Nonlinear membrane theory | XVIII |
| Lichnerowicz, A. | Méthodes tensorielles en mécanique | VIII |
| Lighthill, J. | Aquatic animal locomotion | XIII |
| Lighthill, J. | Typhoons, hurricanes and fluid mechanics | XIX |
| Liñán, A. | Diffusion controlled combustion | XX |
| Linden, P.F. | The fluid mechanics of natural ventilation | XIX |

| Name | Title | Congress |
|---|---|---|
| Lohse, D. | Bubbles in micro- and nano-fluidics | XXII |
| Magnus, K. | Drehbewegungen starrer Körper im zentralen Schwerefeld | XI |
| Martin, J.B. | Computational aspects of integration along the path of loading in elastic-plastic problems | XVIII |
| Matsumoto, Y. | Toward the multi-scale simulation for a human body using the next-generation super-computer | XXIII |
| McIntyre, M.E. | On the role of wave propagation and wave breaking in atmosphere-ocean dynamics | XVIII |
| Meissner, E. | Elastische Oberflächen-Querwellen | II |
| Mettler, E. | Erzwungene nichtlineare Schwingungen elastischer Körper | IX |
| Michel, E. | Raumakustik | III |
| Miles, J.W. | Waves and wave drag in stratified flows | XII |
| Miles, J.W. | The Pendulum, from Huygens' *Horologium* to symmetry breaking and chaos | XVII |
| Milton, G.W. | Composites: a myriad of microstructure independent relations | XIX |
| Moffatt, H.K. | Local and global perspectives in fluid dynamics | XX |
| Moin, P. | The mechanics and prediction of wall-turbulence | XXIII |
| Molinari, A. | Dynamic damage, strain localization and failure of ductile materials | XXIII |
| Monkewitz, P.A. | Flow instabilities and transition in spatially inhomogeneous systems | XX |
| Murakami, S. | Constitutive modeling and analysis of creep, damage, and creep crack growth under neutron irradiation | XVIII |
| Narasimha, R. | Down-to-earth temperatures: the mechanics of the thermal environment | XIX |
| Needleman, A. | Cohesive surface modeling | XXIII |
| Neishtadt, A. | Probability phenomena in perturbated dynamical systems | XXI |
| Nemat-Nasser, S. | Plasticity: inelastic flow of heterogeneous solids at finite strains and rotations | XIX |

| Name | Title | Congress |
|------|-------|----------|
| Nguyen, Q.S. | Stability and bifurcation in dissipative media | XVIII |
| Novozhilov, V.V. | Perspective in phenomenological approach to the problem of fracture | XIII |
| Odqvist, F.K.G. | Non-linear solid mechanics, past, present, and future | XII |
| Ogden, R.W. | Mechanics of rubberlike solids | XXI |
| Oseen, C.W. | Das Turbulenzproblem | III |
| Païdoussis, M.P. | Fluid-structure interactions between axial flows and slender structures | XIX |
| Pouliquen, O. | From dry granular flows to submarine avalanches | XXII |
| Palmov, V.A. | Stationary waves in elasto-plastic and visco-plastic bodies | XX |
| Panetti, M. | Notizie generali sulle oscillazioni dei veicoli | III |
| Pérès, J. | Les méthodes d'analogie en mécanique appliquée | V |
| Peskin, C.S. | Muscle and blood: Fluid dynamics of the heart and its valves | XX |
| Petryk, H. | Instability of plastic deformation processes | XIX |
| Pfeiffer, F. | Multibody dynamics with multiple unilateral contacts | XIX |
| Pouquet, A. | Kolmogorov-like behavior and inertial range vortex dynamics in turbulent compressible and MHD flows | XIX |
| Prager, W. | Recent developments in the mathematical theory of plasticity | VII |
| Prandtl, L. | Spannungsverteilung in plastischen Körpern | I |
| Prandtl, L. | Über die ausgebildete Turbulenz | II |
| Preumont, A. | Some issues in active vibration control of smart structures | XXI |
| Prosperetti, A. | Bubble mechanics: luminescence, noise, and two-phase flow | XVIII |
| Quere, D. | Capillary constructions | XXII |
| Reissner, E. | On the foundations of the theory of elastic shells | XI |
| Rhines, P.B. | Ocean circulation and its influence on climate | XXI |

| Name | Title | Congress |
|---|---|---|
| Rice, J.R. | New perspectives on crack and fault dynamics | XX |
| Rocard, Y. | (title unknown) | VI |
| Roshko, A. | Instability and turbulence in shear flow | XVIII |
| Rudnicki, J. | Failure of rocks in the laboratory and the earth | XXII |
| Sackmann, E. | Microviscoelasticity of cells: cells as viscoplastic bodies | XXI |
| Savage, S.B. | Flow of granular materials | XVII |
| Saville, D.A. | Electrokinetics & electrohydrodynamics in microfluids | XXI |
| Sayir, M.B. | Wave propagation in non-isotropic structures | XVIII |
| Schmidt, E. | Wärmeübergang | IV |
| Schrefler, B.A. | Mechanics of saturated-unsaturated porous materials and quantitative solutions | XIX |
| Sedov, L.I. | Some problems of designing new models of continuum media | XI |
| Shaw, N. | The physical structure of the atmosphere regarded from the dynamical point of view | I |
| Sigmund, O. | Optimum design of MicroElectroMechanical Systems (MEMS) | XX |
| Smith, F.T. | Interactions in boundary-layer transition | XVII |
| Sobczyk, K. | Stochastic modeling of fatigue accumulation | XVII |
| Sobczyk, K. | Stochastic dynamics of engineering systems | XXI |
| Sørensen, J.N. | The aerodynamics of wind turbines | XXII |
| Sottos, N. | Self-healing materials systems: where mechanics meets chemistry | XXII |
| Spiegel, E. | Problems in astrophysical fluid dynamics | XXI |
| Sreenivasan, K. | Cool stuff at cold temperatures | XXIII |
| Sreenivasan, K. | Self-similar multiplier distributions and multiplicative models for energy dissipation in high-Reynolds-number turbulence | XVIII |
| Stein, E. | Error controlled adaptivity for hierarchical models and finite element approximations in structural mechanics | XX |

| Name | Title | Congress |
|---|---|---|
| Sternberg, E. | On singular problems in linearized and finite elastostatics | XV |
| Storakers, B. | Non-linear aspects of delamination in structural members | XVII |
| Stuart, J.T. | Non-linear effects in hydrodynamic stability | X |
| Suquet, P. | Mechanics of polycrystalline and heterogeneous materials at different scales | XXIII |
| Suresh, S. | Nanomechanics and micromechanics of thin films, graded coatings and mechanical/non-mechanical system | XX |
| Sverdrup, H.U. | Oceanic circulation | V |
| Swinney, H.L. | Scaling in quasi-2D turbulence experiments in a rotating flow | XXI |
| Taylor, G.I. | Experiments with rotating fluids | I |
| Taylor, G.I. | The distortion of single crystals of metals | II |
| Taylor, G.I. | The strength of crystals of pure metals and of rock salt | IV |
| Taylor, G.I. | Some recent developments in the study of turbulence | V |
| Taylor, G.I. | (title unknown) | VI |
| Taylor, G.I. | Hydrodynamic theory of detonating explosives | VIII |
| Timoshenko. S. | Stability and strength of thin-walled constructions | III |
| Tollmien, W. | Über Schwingungen in laminaren Strömungen und die Theorie der Turbulenz | VIII |
| van der Giessen, E. | Plasticity in the 21$^{st}$ century | XX |
| van der Meulen, J.H.J | Some physical aspects associated with cavitation | XVII |
| van Steenhoven, A. | Cardiovascular fluid mechanics | XVIII |
| van Wijngaarden, L. | Interplay between air and water | XXI |
| Vekua, I.N. | New methods in mathematical shell theory | XI |
| von Karman, T. | Über die Stabilität der Laminarströmung und die Theorie der Turbulenz | I |
| von Karman, T. | Über elastische Grenzzustände | II |

| Name | Title | Congress |
|---|---|---|
| von Karman, T. | Contributions to the theory of wave-resistance | IV |
| von Karman, T. | Some aspects of the turbulence problem | IV |
| von Mises, R. | Über die bisherigen Ansätze in der klassischen Mechanik der Kontinua | III |
| von Neumann, J. | Automatic computation and hydrodynamics | VIII |
| Wagner, H. | Über das Gleiten von Körpern auf der Wasseroberfläche | IV |
| Wang, R. | Some problems of mechanics in tectonic analysis | XVII |
| Wieghardt, K. | Ship hydrodynamics | XIII |
| Wigley, W.C.S. | Ship wave resistance | III |
| Worster, G. | Dynamics of marine ice sheets | XXIII |
| Wriggers, P. | Characterization of heterogeneous materials by multi-scale simulations | XXII |
| Wright, P. | The mechanics of manufacturing processes | XIX |
| Yamagata, T. | Indian ocean dipole and its possible link with climate modes | XXII |
| Yang, W. | Nanomechanics of graphenes and nano-crystals | XXIII |
| Zaleski, S. | Direct numerical simulation of multiphase flows with volume of fluid methods | XXIII |
| Ziegler, H. | Thermodynamik der Deformationen | XI |
| Zierep, J. | Trends in transonic research | XVIII |

# Appendix 8
# Number of Congress Presentations and their Publication Modes

| | Opening & Closing Lectures | Keynotes & General Lectures | Sectional Papers | Contributed Paper | Contributed Paper | Poster Presentation | Total Presentation | Pages (Numbered & Introd.) | Volumes (Numbered) |
|---|---|---|---|---|---|---|---|---|---|
| 0 Innsbruck 1922 | | | 24 | | | | 24 | 250 | 1 |
| I Delft 1924 | 2 F | 13 | 43 F | | | | 58 | XXII 460 | 1 |
| II Zurich 1926 | 1 F | 10 F | 79 F | | | | 90 | XII 546 | 1 |
| III Stockholm 1930 | | 11 F | 155 F | | | | 166 | XXVIII 1288 | 3 |
| IV Cambridge UK 1934 | | 5 F | 207 A | | | | 212 | XVIII 283 | 1 |
| V Cambridge USA 1938 | 2 F | 3 F | 132 F | | | | 137 | XXXI 748 | 1 |
| VI Paris 1946 | | | | | | | 251 | | 0 |
| VII London 1948 | 1 F | 4 F | 211 F+A | | | | 216 | 1868 | 6 |
| VIII Istanbul 1952 | 3 F | 7 F | 10 F | 336 A | | | 356 | XII 732 | 2 |
| IX Bruxelles 1956 | | 3 F | | 508 F | | | 511 | 4001 | 8 |
| X Stresa 1960 | 4 F&T | 4 F | | 196 A | | | 204 | XXIII 370 | 1 |
| XI Munich 1964 | 2 F&A | 9 | | 143 F | | | 154 | XXVIII 1190 | 1 |
| XII Stanford 1968 | 1* F | 4 F | | | 26 F, 253 T | | 284 | XXIV 420 | 1 |
| XIII Moscow 1972 | 5 F | 5 F | 17 F | 225 T | | | 252 | VIII 366 | 1 |
| XIV Delft 1976 | | 5 F | 23 F | 227 T | | | 256 | X 491 | 2 |
| XV Toronto 1980 | 2 F | 4 F | 34 F | 313 T | | | 353 | XXXI 457 | 1 |
| XVI Lyngby 1984 | 2 F | 2 F | 21 F | 235 T | | 183 | 443 | XXXIII 435 | 1 |
| XVII Grenoble | 2 F | | 13 F | 435 | | | 450 | XXXIX | 1 |

© The Author(s) 2016
P. Eberhard and S. Juhasz (eds.), *IUTAM*,
DOI 10.1007/978-3-319-31063-3

Appendix 8: Number of Congress Presentations and their Publication Modes

| | | | | | | | | | |
|---|---|---|---|---|---|---|---|---|---|
| 1988 | | | | T | | | | 453 | |
| XVIII Haifa 1992 | 2 F | | 15 F | 232T | | 171 T | 420 | XXXII 459 | 1 |
| XIX Kyoto 1996 | 2 F | | 12F, 3 T | 442T | | 244 T | 703 | XXXIX 623 | 1 |
| XX Chicago 2000 | 2 F | | 12 F | 704T | | 379 T | 1126 | LXXX 582 | 1 |
| XXI Warsaw 2004 | 2 F | | 18 F | 1253 T | | | 1273 | LXXIII 421 | 1 |
| XXII Adelaide 2008 | 2 F | | 6 F, 11 T | 921T | | | 940 | CXXII 305 | 1 |
| XXIII Beijing 2012 | 2 F | | 17 F | 1252 T | | | 1271 | E | E |
| **total** | | | | | | | **10150** | **16748** | **38** |

F = fully published,    A = published by abstract,    T = published by title

E = published as e-book

# Appendix 9
# Congress Papers by Subject (1922–1984)

| | Foundations and Basic Methods | Vibration Dynamics | Control | Solids | Fluids | Thermal Sciences | Earth Sciences | Energy Systems & Env. | Biosciences | Other | Total |
|---|---|---|---|---|---|---|---|---|---|---|---|
| 0 Innsbruck, 1922 | 0 | 2 | 0 | 0 | 21 | 0 | 1 | 0 | 0 | 0 | 24 |
| I Delft, 1924 | 5 | 3 | 0 | 25 | 22 | 0 | 3 | 0 | 0 | 0 | 58 |
| II Zurich, 1926 | 2 | 16 | 0 | 45 | 24 | 2 | 1 | 0 | 0 | 0 | 90 |
| III Stockholm, 1930 | 5 | 44 | 0 | 59 | 56 | 1 | 1 | 0 | 0 | 0 | 166 |
| IV Cambridge UK, 1934 | 5 | 20 | 0 | 72 | 86 | 28 | 1 | 0 | 0 | 0 | 212 |
| V Cambridge USA, 1938 | 4 | 23 | 0 | 47 | 56 | 4 | 3 | 0 | 0 | 0 | 137 |
| VI Paris, 1946 | 7 | 25 | 0 | 82 | 109 | 24 | 0 | 0 | 0 | 4 | 251 |
| VII London, 1948 | 6 | 23 | 1 | 75 | 78 | 24 | 0 | 0 | 0 | 9 | 216 |
| VIII Istanbul, 1952 | 37 | 33 | 2 | 118 | 129 | 29 | 1 | 0 | 0 | 7 | 356 |
| IX Brussels, 1956 | 73 | 52 | 0 | 164 | 210 | 4 | 8 | 0 | 0 | 0 | 511 |
| X Stresa, 1960 | 2 | 29 | 0 | 78 | 88 | 6 | 0 | 0 | 0 | 1 | 204 |
| XI Munich, 1964 | 2 | 20 | 0 | 59 | 58 | 6 | 1 | 0 | 1 | 7 | 154 |
| XII Stanford, 1968 | 7 | 37 | 1 | 134 | 78 | 7 | 3 | 0 | 7 | 10 | 284 |
| XIII Moscow, 1972 | 8 | 43 | 1 | 84 | 93 | 5 | 2 | 0 | 7 | 9 | 252 |
| XIV Delft, 1976 | 16 | 31 | 1 | 103 | 74 | 10 | 5 | 0 | 7 | 9 | 256 |
| XV Toronto, 1980 | 20 | 58 | 1 | 146 | 100 | 14 | 2 | 1 | 1 | 10 | 353 |
| XVI Lyngby, 1984 | 39 | 74 | 2 | 165 | 117 | 26 | 12 | 4 | 4 | 0 | 443 |
| total no. | 238 | 533 | 9 | 1456 | 1399 | 190 | 44 | 5 | 27 | 66 | 3967 |
| total % | 6.0 | 13.4 | 0.23 | 36.7 | 35.3 | 4.8 | 1.1 | 0.12 | 0.7 | 1.7 | 100 |

General Lectures, Sectional Lectures, and Contributed Papers are counted

© The Author(s) 2016
P. Eberhard and S. Juhasz (eds.), *IUTAM*,
DOI 10.1007/978-3-319-31063-3

# Appendix 10
# Bibliographic Details Congress Proceedings

$0^{th}$ Conference, Innsbruck (Austria), 10–13 September 1922. Vorträge aus dem Gebiete der Hydro- und Aerodynamik (Innsbruck 1922), edited by T.V. Karman und T. Levi-Civita, Verlag von Julius Springer, Berlin, 1924.

$1^{st}$ Congress, Delft (Netherlands), 22–26 April 1924. Proceedings of the First International Congress for Applied Mechanics, Delft 1924, edited by C.B. Biezeno and J.M. Burgers. Technische Boekhandel en Drukkerij J. Waltman Jr., Delft, 1925.

$2^{nd}$ Congress, Zürich (Switzerland), 12–17 September 1926. Verhandlungen—Comptes rendus—Proceedings of the 2nd International Congress for Applied Mechanics, Zürich, 12–17 September 1926, edited by E. Meissner. Orell Füssli Verlag, Zürich und Leipzig, 1927.

$3^{rd}$ Congress, Stockholm (Sweden), 24–29 August 1930. Verhandlungen—Compte rendus—Proceedings of the 3rd International Congress for Applied Mechanics, edited by A.C.W. Oseen und W. Weibull (3 vol.). AB. Sveriges Litografiska Tryckerier, Stockholm, 1931.

$4^{th}$ Congress, Cambridge (UK), 3–9 July 1934. Proceedings of the Fourth International Congress for Applied Mechanics, Cambridge, UK, 3–9 July, 1934. University Press, Cambridge, 1935.

$5^{th}$ Congress, Cambridge (Massachusetts, USA), 12–16 September 1938. Proceedings of the Fifth International Congress for Applied Mechanics, held at Harvard University and the Massachusetts Institute of Technology, Cambridge, Massachusetts, September 12–16, 1938, edited by J.P. den Hartog and H. Peters, John Wiley and Sons. New York (USA), and Chapman and Hall. London (UK), 1939.

$6^{th}$ Congress, Paris (France), 22–29 September 1946. Proceedings not published (was given in the hands of Gauthier-Villars, Paris).

$7^{th}$ Congress, London (UK), 5–11 September 1948. Proceedings of the Seventh International Congress for Applied Mechanics, 1948, published by the

© The Author(s) 2016
P. Eberhard and S. Juhasz (eds.), *IUTAM*,
DOI 10.1007/978-3-319-31063-3

Organizing Committee (Introduction, Vol. I, Vol. II—Parts 1 and 2, Vol. III, Vol. IV).

8th Congress, Istanbul (Turkey), 20–28 August 1952. Proceedings of the Eighth International Congress on Theoretical and Applied Mechanics, edited by F. Rolla and W.T. Koiter, published by the Organizing Committee (Vol. I, Vol. II). Faculty of Sciences, University of Istanbul, Turkey, 1953.

9th Congress, Brussels (Belgium), 5–13 September 1956. Proceedings published by the Organizing Committee (Vol. I to Vol. VIII). Free University of Brussels, Brussels, 1957.

10th Congress, Stresa (Italy), 31 August–7 September 1960. Proceedings of the Tenth International Congress of Applied Mechanics, published by the Consiglio Nazionale delle Ricerche, Roma (Italia), printed by Elsevier Publishing Company, Amsterdam New York, 1962.

11th International Congress on Theoretical and Applied Mechanics (ICTAM), Munich (Germany), 30 August–5 September 1964. The Proceedings, edited by H. Görtler, have been published by Springer-Verlag, Berlin, 1966.

12th International Congress on Theoretical and Applied Mechanics (ICTAM), Stanford, Cal. (USA), 26–31 August 1968. The Proceedings, edited by M. Hetényi and W.G. Vincenti, have been published by Springer-Verlag, Berlin, 1969.

13th International Congress on Theoretical and Applied Mechanics (ICTAM), Moscow (USSR), 21–26 August 1972. The Proceedings, edited by E. Becker and G.K. Mikhailov, have been published by Springer-Verlag, Berlin, 1973.

14th International Congress on Theoretical and Applied Mechanics (ICTAM), Delft (Netherlands), 30 August–4 September 1976. The Proceedings, edited by W.T. Koiter, have been published by North-Holland Publishing Company, Amsterdam New York Oxford, 1977.

15th International Congress on Theoretical and Applied Mechanics (ICTAM), Toronto (Canada), 17–23 August 1980. The Proceedings, edited by F.P. J. Rimrott and B. Tabarrok, have been published by North-Holland Publishing Company, Amsterdam New York Oxford 1980.

16th International Congress on Theoretical and Applied Mechanics (ICTAM), Lyngby (Denmark), 19–25 August 1984. The Proceedings, edited by F.I. Niordson and N. Olhoff, have been published by North-Holland Publishing Company, Amsterdam, 1985.

17th International Congress on Theoretical and Applied Mechanics (ICTAM), Grenoble (France), 21–27 August 1988. The Proceedings, edited by P. Germain, M. Piau and D. Caillerie, have been published by North-Holland, Elsevier Science Publishers, Amsterdam, 1989. ISBN 0-444-87302-3.

18th International Congress on Theoretical and Applied Mechanics (ICTAM), Haifa (Israel), 22–28 August 1992. The Proceedings, edited by S.R. Bodner, J. Singer, A. Solan and Z. Hashin, have been published by Elsevier Science Publ., Amsterdam, 1993. ISBN 0-444-88889-6.

19th International Congress on Theoretical and Applied Mechanics (ICTAM), Kyoto (Japan), 25–31 August 1996. The Proceedings, edited by T. Tatsumi, E. Watanabe, T. Kambe, have been published by Elsevier Science Publishers, Amsterdam, 1997. ISBN 0-444-82446-4.

20th International Congress on Theoretical and Applied Mechanics (ICTAM), Chicago (USA), 27 August–2 September 2000. The Proceedings, entitled "Mechanics for a new Millenium",edited by H. Aref and J.W. Phillips, have been published by Kluwer Academic Publishers, Dordrecht, 2001. ISBN 0-7923-7156-9.

21[th] International Congress on Theoretical and Applied Mechanics (ICTAM), Warsaw (Poland), 15–21 August 2004. The Proceedings, entitled "Mechanics of the 21st Century", edited by W. Gutkowski and T.A. Kowaleski, have been published by Springer, Dordrecht, 2005. ISBN 1-4020-3456-3.

22nd International Congress on Theoretical and Applied Mechanics (ICTAM), Adelaide (Australia), 24–29 August 2008. The Proceedings, entitled "Mechanics Down Under", edited by J. Denier and M. Finn, have been published by Springer, Dordrecht, 2013, both as an eBook (ISBN 978-94-007-5968-8) and as a Hardcover (ISBN 978-94-007-5967-1).

23rd International Congress on Theoretical and Applied Mechanics (ICTAM), Beijing (China), 19–24 August 2012. The Proceedings, entitled "Mechanics for the World", edited by Y. Bai, J. Wang and D. Fang, have been published by Elsevier, 2014, IUTAM e-Procedia series (Procedia IUTAM Volume 10).

# Appendix 11
# Conversazione

**V Congress, Cambridge, Massachusetts 1938**

"Tuesday evening there was a Conversazione in the laboratories of the Mechanical Engineering Department of the Massachusetts Institute of Technology. Such an event was on the program of the Fourth Congress and an attempt was made to repeat it. Exhibits of apparatus were made by:

- Determination of Grain Orientation in Steel by Electromagnetic Methods: Professor Bitter
- Wear Testing Machine: Professor E. Buckingham
- Various Instruments for Measuring Dynamic Stresses: Professor A.V. de Forest
- Vibration Measuring Equipment: Professor C.S. Draper
- Model Pelton Wheel: Professor J.J. Eames
- Stroboscopic Exhibits: Professor H.E. Edgerton
- Instrument for Measuring the Damping Capacity of Material: Dr. H.E. Hall
- Very Precisely Graduated Scales: Professor G.R. Harrison
- The Cimema Integraph: Professor H.L. Hazen and Dr. G.S. Brown
- Exhibit of Nitrided Material: Professor V.O. Homerberg
- Museum of Naval Architecture and Marine Engineering: Professor J.R. Jack
- Demonstration of Solidification: Professor P.E. Kyle
- Equipment for the Study of Marine Propellers: Professor P.M. Lewis
- Plastic Deformation as Revealed by New Plastiscope: Professor C.W. MacGregor
- Apparatus for Combined Stress Tests: Professors C.W. MacGregor and J.M. Lessells
- Exhibit of Photoelasticity: Dr. W.M. Murray
- Chart of Creep Data: Professor F.H. Norton
- Cavitation Apparatus in Operation: Professor H. Peters
- Model Vortex: Professor K.C. Reynolds
- Model of a Structure Subjected to an Earthquake: Professor A.C. Ruge
- Textiles and Textile Testing: Professor E.R. Schwarz
- Seismograph Used to Record Disturbances Due to Large Quarry Blasts: L.B. Slichter

© The Author(s) 2016
P. Eberhard and S. Juhasz (eds.), *IUTAM*,
DOI 10.1007/978-3-319-31063-3

- High Speed Engine Indicator: Professor E.S. Taylor
- Structures Laboratory: Professor J.B. Wilbur
- Demonstration of the Shielding and Reradiating Properties of Aluminum Foil: Professor G.B. Wilkes
- The Corrosion of Steel in Salt Water: Professor J.C.G. Wulff
- Model Illustrating Method of Stress Analysis in Dirigibles: Dr. K. Arnstein, Goodyear-Zeppelin Corporation
- The Harvard Radio-Meteorograph: Professor C.F. Brooks, Blue Hill Observatory
- Exhibit of Three-Dimensional Photoelasticity: Dr. M. Hetenyi, Westinghouse Research Laboratories
- Exhibit of Photoelasticity: Professor M.M. Frocht, Carnegie Institute of Technology
- Pneumatic Extensometer and Pneumatic Surface Quality Meter: Mr. H. de Leiris and Mr. P. Nicolau, Service Technique des Constructions Navales
- Apparatus for Measuring Stresses Caused by Vibration: Dr. R.K. Mueller, Hamilton Standard Propellers
- Model Demonstrating the Creep of a Rotating Shaft: Dr. A. Nadai, Westinghouse Research Laboratories
- Model Wind Tunnel: Professor J.R.Weske. Case School of Applied Science

# Appendix 12
# Symposia Locations

Symposia from 1949 to 1985

| Country | City | Frequency | Frequency/Country |
|---|---|---|---|
| Australia | Canberra | 2 | 3 |
| | Margaret River | 1 | |
| Austria | Vienna | 3 | 4 |
| | Innsbruck | 1 | |
| Belgium | Bruxelles | 3 | 5 |
| | Liège | 1 | |
| | Louvain-la-Nueve | 1 | |
| Brazil | Rio de Janeiro | 1 | 2 |
| | Sao Paulo | 1 | |
| CSSR | Liblice | 2 | 3 |
| | Prague | 1 | |
| Canada | Calgary | 1 | 6 |
| | Ottawa | 3 | |
| | Quebec | 1 | |
| | Waterloo | 1 | |
| Denmark | Copenhagen | 2 | 3 |
| | Lyngby | 1 | |
| France | Dourdan | 1 | 22 |
| | Grenoble | 1 | |
| | Lyon | 1 | |
| | Marseilles | 3 | |
| | Nancy | 1 | |
| | Nice | 1 | |
| | Palaiseau | 1 | |
| | Paris | 1 | |
| | Poitiers | 1 | |
| | Senlis | 2 | |
| | Toulouse | 8 | |
| | Porquerolles | 1 | |
| Germany / FRG | Aachen | 1 | 15 |
| | Berlin-West | 1 | |
| | Essen | 1 | |
| | Freiburg i.Br. | 1 | |
| | Göttingen | 2 | |
| | Karlsruhe | 3 | |
| | Munich | 1 | |
| | Nuremberg | 1 | |
| | Paderborn | 1 | |
| | Stuttgart | 2 | |
| | Tutzing | 1 | |
| Germany/GDR | Frankfurt / Oder | 1 | 1 |
| India | Delhi | 1 | 1 |
| Israel | Haifa | 1 | 2 |
| | Tel Aviv | 1 | |
| Italy | Pallanza | 1 | 4 |
| | Torino | 1 | |
| | Udine | | |
| | Varenna | | |

© The Author(s) 2016
P. Eberhard and S. Juhasz (eds.), *IUTAM*,
DOI 10.1007/978-3-319-31063-3

| Country | City | Frequency | Frequency/Country |
|---|---|---|---|
| Japan | Kyoto | 2 | 6 |
| | Tokyo | 4 | |
| Netherlands | Delft | 5 | 6 |
| | Enschede | 1 | |
| New Zealand | Hamilton | 1 | 1 |
| Poland | Tuzcno | 1 | 4 |
| | Warsaw | 3 | |
| Portugal | Lisbon | 1 | 1 |
| Spain | Madrid | 1 | 1 |
| Sweden | Gothenburg | 2 | 4 |
| | Stockholm | 2 | |
| Switzerland | Celerina | 1 | 1 |
| UK | Cambridge | 5 | 16 |
| | Coventry | 1 | |
| | East Kilbride | 1 | |
| | Lancaster | 1 | |
| | London | 3 | |
| | Newcastle Upon Tyne | 1 | |
| | Oxford | 1 | |
| | Reading | 1 | |
| | Sheffield | 1 | |
| | Southampton | 1 | |
| USA | Ann Arbor, MI | 1 | 19 |
| | Bethlehem, PA | 1 | |
| | Blacksburg, VI | 1 | |
| | Cambridge, MA | 3 | |
| | Charlottesville, VI | 1 | |
| | Evanston, IL | 3 | |
| | Ithaca, NY | 1 | |
| | La Jolla, CA | 1 | |
| | Monterey, CA | 1 | |
| | Pasadena, CA | 1 | |
| | Providence, RI | 1 | |
| | Stanford, CA | 1 | |
| | Washington, DC | 2 | |
| | Hershey, PA | 1 | |
| USSR | Kiev | 1 | 8 |
| | Leningrad | 1 | |
| | Novosibirsk | 2 | |
| | Tallinn | 1 | |
| | Tbilisi | 2 | |
| | Yalta | 1 | |
| Yugoslavia | Dubrovnik | 1 | 1 |

## Symposia from 1986 to 2015

| Country | City | Frequency | Frequency/Country |
|---|---|---|---|
| Australia | Brisbane | 1 | 6 |
| | Broome | 1 | |
| | Melbourne | 1 | |
| | Sydney | 3 | |
| Austria | Graz | 1 | 4 |
| | Innsbruck | 1 | |
| | Vienna | 2 | |
| Belgium | Leuven | 1 | 1 |
| Brazil | Petropolis | 1 | 1 |
| Canada | Victoria, British Columbia | 1 | 4 |
| | St. John's, Newfoundland | 1 | |
| | Kingston, Ontario | 1 | |
| | Waterloo | 1 | |
| China | Beijing | 7 | 15 |
| | Hangzhou | 1 | |
| | Hong Kong | 3 | |
| | Lanzhou | 1 | |
| | Nanjing | 1 | |
| | Shanghai | 1 | |
| | Xi'an | 1 | |
| China / Taipei | Tainan | 1 | 4 |
| | Taipei | 3 | |
| Cyprus | Limassol | 1 | 1 |
| Denmark | Aalborg | 2 | 9 |
| | Copenhagen | 2 | |
| | Lyngby | 4 | |
| | Rungstegaard | 1 | |
| Estonia | Tallinn | 1 | 1 |
| Egypt | Cairo | 1 | 1 |
| France | Cachan | 2 | 22 |
| | Compiegne | 1 | |
| | Grenoble | 1 | |
| | Lyon | 1 | |
| | Marseille | 3 | |
| | Metz | 1 | |
| | Nice | 1 | |
| | Paris | 5 | |
| | Poitiers | 3 | |
| | Sophia-Antipolis | 1 | |
| | Toulouse | 2 | |
| | Villard-de-Lans | 1 | |
| Germany | Aachen | 1 | 31 |
| | Attendorn | 1 | |
| | Bochum | 3 | |
| | Bremen | 1 | |
| | Dresden | 1 | |
| | Erlangen | 1 | |
| | Frankfurt | 1 | |
| | Freiberg | 1 | |
| | Göttingen | 6 | |
| | Hamburg | 1 | |
| | Hannover | 3 | |
| | Karlsruhe | 1 | |
| | Magdeburg | 1 | |

| Country | City | Frequency | Frequency/Country |
|---|---|---|---|
| | Munich | 2 | |
| | Stuttgart | 6 | |
| | Wuppertal | 1 | |
| Greece | Chania | 1 | 3 |
| | Corfu | 1 | |
| | Santorini Island | 1 | |
| Hungary | Budapest | 1 | 1 |
| India | Bangalore | 6 | 11 |
| | Goa | 1 | |
| | Hyderabad | 1 | |
| | Kanpur | 1 | |
| | Madras | 1 | |
| | New Delhi | 1 | |
| Ireland | Dublin | 4 | 6 |
| | Galway | 1 | |
| | Limerick | 1 | |
| Israel | Haifa | 2 | 2 |
| Italy | Brescia | 1 | 6 |
| | Capri | 1 | |
| | Reggio Calabria | 1 | |
| | Rome | 1 | |
| | Torino | 2 | |
| Japan | Fukuoka | 1 | 15 |
| | Hayama | 1 | |
| | Kyoto | 3 | |
| | Nagoya | 1 | |
| | Noda | 1 | |
| | Osaka | 1 | |
| | Sendai | 1 | |
| | Tokyo | 1 | |
| | Yamagata | 1 | |
| Lavia / USSR | Riga | 2 | 2 |
| Morocco | Marrakesch | 1 | 1 |
| Mexico | Guanajuato | 1 | 2 |
| | Oaxaca | 1 | |
| Netherlands | Delft | 1 | 5 |
| | Eindhoven | 2 | |
| | Kerkrade | 2 | |
| Norway | Trondheim | 1 | 1 |
| Poland | Cracow | 3 | 6 |
| | Kazimierz Dolny | 1 | |
| | Zakopane | 2 | |
| Russia / USSR | Izhersk | 1 | 9 |
| | Novosibirsk | 2 | |
| | Moscow | 5 | |
| | St. Petersburg | 1 | |
| South Africa | Cape Town | 1 | 1 |
| Spain | Sevilla | 1 | 1 |
| Sweden | Kiruna | 1 | 4 |
| | Luleå | 1 | |
| | Stockholm | 2 | |
| Switzerland | Zürich | 2 | 2 |
| Tobago | | 1 | 1 |
| Turkey | Istanbul | 1 | 1 |
| UK | Aberdeen | 1 | 26 |
| | Bath | 1 | |

| Country | City | Frequency | Frequency/Country |
|---------|------|-----------|-------------------|
|  | Birmingham | 2 |  |
|  | Cambridge | 7 |  |
|  | Cardiff | 3 |  |
|  | Liverpool | 1 |  |
|  | London | 2 |  |
|  | Manchester | 3 |  |
|  | Nottingham | 1 |  |
|  | Southampton | 2 |  |
|  | Uxbridge | 1 |  |
|  | Warwick / Coventry | 2 |  |
| USA | Austin, TX | 4 | 32 |
|  | Blacksburg, VA | 1 |  |
|  | Boulder, CO | 3 |  |
|  | Cape Cod, MA | 1 |  |
|  | Cape May, NJ | 1 |  |
|  | Estes Park, CO | 1 |  |
|  | Evanston, IL | 1 |  |
|  | Gainesville, FL | 1 |  |
|  | Ithaca, NY | 1 |  |
|  | La Jolla, CA | 1 |  |
|  | Notre Dame, IN | 1 |  |
|  | Pasadena, CA | 1 |  |
|  | Pensacola, FL | 1 |  |
|  | Postdam, NY | 1 |  |
|  | Princeton, NJ | 1 |  |
|  | Providence, RI | 1 |  |
|  | Rutgers, NJ | 1 |  |
|  | San Antonio, TX | 2 |  |
|  | Sedona, AZ | 1 |  |
|  | Stanford, CA | 3 |  |
|  | Tbilisi, GA | 1 |  |
|  | Troy, NY | 1 |  |
|  | Urbana - Champaign, IL | 2 |  |
| Vietnam | Hanoi | 1 | 1 |

1949-1985:

| | |
|---|---|
| total number of symposia: | 139 |
| number of cities: | 91 |
| number of countries: | 25 |

1986-2015:

| | |
|---|---|
| total number of symposia: | 239 |
| number of cities: | 135 |
| number of countries: | 37 |

1949-2015:

| | |
|---|---|
| total number of symposia: | 378 |
| total number of countries: | 41 |

# Appendix 13
# Symposia by Subject, Location, and Number of Participants

| No. | Date | Co-Sponsor | Symposia Title | Sub-ject | Location City, Country | Partici-pants | Coun-tries |
|---|---|---|---|---|---|---|---|
| 49-1 | 08/16/49 | IAU | Problems on Motion of Gaseous Masses of Cosmical Dimen-sions, 1st | F | Paris, France | 51 | |
| 50-1 | 06/23/50 | | Colloque de Mécanique | M | Pallanza, Italy | 27 | 11 |
| 50 2 | 09/12/50 | IUGG | Plastic Flow and Deformation within the Earth | S | Hershey, PA, USA | 23 | 7 |
| 51-1 | 09/18/51 | IRSI | Non-Linear Vibrations | D | Ile de Porquerolles, France | 43 | 14 |
| 52-1 | 04/25/52 | | Colloques sur les Zones Andes | O | Bruxelles, Bel-gium | 28 | 14 |
| 53-1 | 07/06/53 | IAU | Gas Dynamics of Interstellar Clouds, 2nd | F | Cambridge, UK | 63 | 9 |
| 54-1 | 07/29/54 | | Photoélasticité et Photoplasticité | S | Bruxelles, Bel-gium | 36 | 11 |
| 55-1 | 05/25/55 | SNCM | Colloquium on Fatigue | S | Stockholm, Sweden | 147 | 16 |
| 55-2 | 09/26/55 | INTA/ UNESCO | Deformation and Flow in Solids | S | Madrid, Spain | 71 | 10 |
| 57-1 | 06/24/57 | IAU | Cosmical Gas Dynamics, 3rd | F | Cambridge, MA, USA | 82 | 8 |
| 57-2 | 08/26/57 | | Couche-Limite | F | Freiburg i.B., Germany | 85 | 17 |
| 58-1 | 08/24/58 | IUGG | Atmospheric Diffusion and Air Pollution | E | Oxford, UK | 100 | 20 |
| 58-2 | 09/02/58 | | Non-Homogénéité en Elasticité et Plasticite | S | Warsaw, Poland | 64 | 14 |
| 59-1 | 07/09/59 | IUGG,IAU, URSI | Fluid Mechanics in the Iono-sphere | F | Ithaca, NY, USA | 60 | |
| 59-2 | 08/24/59 | | Theory of Thin Elastic Shells | S | Delft, Nether-lands | 67 | 13 |
| 60-1 | 01/17/60 | | Magneto-Fluid Dynamics | F | Washington, D.C., USA | 100 | |
| 60-2 | 07/11/60 | NSF | Creep in Structures | S | Stanford, CA, USA | 60 | 9 |

© The Author(s) 2016
P. Eberhard and S. Juhasz (eds.), *IUTAM*,
DOI 10.1007/978-3-319-31063-3

| No. | Date | Co-Sponsor | Symposia Title | Sub-ject | Location City, Country | Partici-pants | Coun-tries |
|---|---|---|---|---|---|---|---|
| 60-3 | 08/18/60 | IAU | Cosmical Gas Dynamics, 4th | F | Varenna, Italy | 63 | 14 |
| 61-1 | 09/04/61 | IUGG | Fundamental Problems in Turbulence and Their Relation to Geophysics | F | Marseille, France | 80 | 15 |
| 61-2 | 09/12161 | | Theory of Non-Linear Vibrations | D | Kiev, USSR | 84 | 15 |
| 62-1 | 04/21/62 | | Second Order Effects in Elasticity, Plasticity and Fluid Dynamics | S | Haifa, Israel | 75 | 14 |
| 62-2 | 10/28/62 | | La Dynamique des Satellites | D | Paris, France | 40 | 9 |
| 62-3 | 08/20/62 | | Gyrodynamics | D | Celerina, Switzerland | 63 | 13 |
| 62-4 | 09/03/62 | | Transsonicum, 1st | F | Aachen, FRG | 102 | 16 |
| 63-1 | 04/03/63 | | Stress Waves in Anelastic Solids | S | Providence, RI, USA | 61 | 9 |
| 63-2 | 09/17/63 | | Applications of the Theory of Functions in Continuum Mechanics | M | Tbilisi, USSR | 124 | 19 |
| 64-1 | 04/01/64 | | Rhéologie et Mécanique des Sols | F | Grenoble, France | 102 | 16 |
| 64-2 | 07/06/64 | | Concentrated Vortex Motions in Fluids | F | Ann Arbor, MI, USA | 150 | 13 |
| 65-1 | 04/13/65 | | Recent Advance in Linear Vibration Mechanics | D | Paris, France | 54 | 7 |
| 65-2 | 04/20/65 | IAU, COSPAR | Trajectories of Artificial Celestial Bodies as Determined From Observation | G | Paris, France | 32 | 10 |
| 65-3 | 09/02/65 | IAU | Cosmical Gas Dynamics, 5th | F | Nice, France | 52 | 12 |
| 66-1 | 03/28/66 | | Rotating Fluid Systems | F | La Jolla, CA, USA | 75 | 10 |
| 66-2 | 06/22/66 | | Irreversible Aspects of Continuum Mechanics | M | Vienna, Austria | 57 | 15 |
| 66-3 | 06/27/66 | | Transfer of Physical Characteristics in Moving Fluids | F | Vienna, Austria | 40 | 9 |
| 66-4 | 09/19/66 | IUGG | Boundary Layers and Turbulence Including Geophysical Applications | F | Kyoto, Japan | 182 | 13 |
| 67-1 | 08/28/67 | | Generalized Cosserat Continuum and the Continuum Theory of Dislocations with Applications | S | Stuttgart, FRG | 70 | 18 |
| 67-2 | 09/05/67 | | Theory of Thin Shells, 2nd | S | Copenhagen, Denmark | 75 | 16 |
| 67-3 | 09/11/67 | | Behaviour of Dense Media Under High Dynamic Pressures | S | Paris, France | 186 | 12 |

| No. | Date | Co-Sponsor | Symposia Title | Sub-ject | Location City, Country | Partici-pants | Coun-tries |
|---|---|---|---|---|---|---|---|
| 68-1 | 06/26/68 | | Thermoinelasticity | T | E. Kilbride, Glasgow, UK | 72 | 15 |
| 68-2 | 08/19/68 | | High-Speed Computing in Fluid Dynamics | F | Monterey, CA, USA | 162 | 13 |
| 69-1 | 03/24/69 | | Flow of Fluid-Solid Mixtures | F | Cambridge, UK | 110 | 19 |
| 69-2 | 03/31/69 | IUPAP | Electrohydrodynamics | F | Cambridge, MA, USA | 156 | 10 |
| 69-3 | 05/20/69 | COSPAR, IAU, IUGG | Dynamics of Satellites | D | Prague, Czecho-slovakia | 117 | 10 |
| 69-4 | 09/08/69 | | Instability of Continuous Systems | S | Karlsruhe, FRG | 99 | 19 |
| 69-5 | 09/08/69 | IAU | Cosmical Gas Dynamics 6th | F | Yalta, USSR | 140 | 13 |
| 70-1 | 08/17/70 | | Creep in Structures 2nd | S | Gothenburg, Sweden | 81 | 18 |
| 70-2 | 08/23/70 | | High-Speed Computing of Elastic Structures | S | Liege, Belgium | 98 | 15 |
| 71-1 | 05/12/71 | IUGG | Flow of Multiphase Fluids in Porous Media | F | Calgary, Canada | 41 | 14 |
| 71-2 | 05/24/71 | | Unsteady Boundary Layers | F | Quebec, Canada | 84 | 13 |
| 71-3 | 06/22/71 | | Non-Steady Flow of Water at High Speeds | F | Leningrad, USSR | 203 | 17 |
| 71-4 | 09/13/71 | | Dynamics of Ionized Gases | F | Tokyo, Japan | 56 | 8 |
| 72-1 | 04/17/72 | ITTC | Directional Stability and Control of Bodies Moving in Water | D | London, UK | 80 | 12 |
| 72-2 | 07/11/72 | | Stability of Stochastic Dynamical Systems | D | Coventry, UK | 40 | 12 |
| 72-3 | 08/14/72 | IAHR | Flow-Induced Structural Vibrations | FS | Karlsruhe, FRG | 210 | 27 |
| 73-1 | 04/08/73 | IUGG | Turbulent Diffusion in Environmental Pollution | E | Charlottesville, VA, USA | 128 | 17 |
| 73-2 | 08/21/73 | | Optimization in Structural Design | OP | Warsaw, Poland | 78 | 21 |
| 73-3 | 09/05/73 | IAU | Stability of the Solar System and of Small Stellar Systems | O | Warsaw, Poland | 100 | 25 |
| 73-4 | 09/09/73 | | Photoelastic Effect and Its Applications | S | Bruxelles, Belgium | 40 | 14 |
| 74-1 | 06/17/74 | | Buckling of Structures | S | Cambridge, MA, USA | 93 | 17 |
| 74-2 | 06/19/74 | COSPAR, IAU | Satellite Dynamics | D | Sao Paulo, Brazil | 50-60 | |
| 74-3 | 08/12/74 | | Dynamics of Rotors | D | Lyngby, Denmark | 87 | 18 |
| 74-4 | 08/20/74 | | Mechanics of the Contact Between Deformable Bodies | S | Enschede, Netherlands | 73 | 15 |
| 74-5 | 09/02/74 | | Mechanics of Visco-Elastic Media and Bodies | S | Gothenburg, Sweden | 55 | 15 |

| No. | Date | Co-Sponsor | Symposia Title | Sub-ject | Location City, Country | Partici-pants | Coun-tries |
|---|---|---|---|---|---|---|---|
| 75-1 | 08/18/75 | | Dynamics of Vehicles on Roads and on Railway Tracks | D | Delft, Nether-lands | 80 | 15 |
| 75-2 | 09/01/75 | IUPAP | Biodynamics of Animal Loco-motion | B | Cambridge, UK | 58 | 14 |
| 75-3 | 09/01/75 | IMU | Applications of Methods of Functional Analysis to Problems of Mechanics | M | Marseilles, France | 132 | 15 |
| 75-4 | 09/08/75 | | Transsonicum, 2nd | F | Göttingen, FRG | 156 | 19 |
| 76-1 | 06/07/76 | | Structure of Turbulence and Drag Reduction | F | Washington, D.C., USA | 115 | 14 |
| 76-2 | 07/19/76 | | Stochastic Problems in Dynam-ics | D | Southhampton, UK | 62 | 15 |
| 76-3 | 07/20/76 | | Surface Gravity Waves in Water of Varying Depth | F | Canberra, Aus-tralia | 45 | 8 |
| 76-4 | 10/18/76 | ATTAG | Aeroelasticity in Turbomachines | D | Paris, France | 85 | 7 |
| 77-1 | 08/24/77 | | High Velocity Deformation of Solids | S | Tokyo, Japan | 91 | 9 |
| 77-2 | 08/29/77 | | Dynamics of Multibody Systems | D | Munich, FRG | 124 | 15 |
| 77-3 | 09/11/77 | | Modern Problems in Elastic Wave Propagation | S | Evanston, IL, USA | 95 | |
| 77-4 | 09/19/77 | | Dynamics of Vehicles on Roads and Tracks | D | Vienna, Austria | 93 | 16 |
| 77-5 | 12/05/77 | IUGG | Monsoon Dynamics | E | Delhi, India | 177 | 17 |
| 78-1 | 05/28/78 | COSPAR, IAU, IUGG | Atmospheres and Surfaces of Plants | E | Innsbruck, Aus-tria | ? | |
| 78-2 | 08/22/78 | | Shell Theory | S | Tbilisi, USSR | 96 | 18 |
| 78-3 | 08/25/77 | IMU | Group Theoretical Methods in Mechanics | M | Novosibirsk, USSR | 65 | 12 |
| 78-4 | 08/28/78 | | Non-Newtonian Fluid Mechan-ics | F | Louvain-la-Neuve, Belgium | 92 | 20 |
| 78-5 | 08/28/78 | | Metal Forming Plasticity | S | Tutzing, FRG | 75 | 15 |
| 78-6 | 09/11/78 | | Variational Methods in the Me-chanics of Solids | S | Evanston, IL, USA | 120 | 22 |
| 79-1 | 06/04/79 | | Structural Control | D | Waterloo, Cana-da | 45 | 10 |
| 79-2 | 08/06/79 | | Physics and Mechanics of Ice | S | Copenhagen, Denmark | 63 | 14 |
| 79-3 | 08/28/79 | ICA, AIAA | Mechanics of Sound Generation in Flows | F | Göttingen, FRG | 82 | 12 |
| 79-4 | 09/03/79 | IAHR | Practical Experiences with Flow-Induced Vibrations | D | Karlsruhe, FRG | 239 | 23 |

| No. | Date | Co-Sponsor | Symposia Title | Sub-ject | Location City, Country | Partici-pants | Coun-tries |
|-----|------|-----------|----------------|----------|------------------------|---------------|------------|
| 79-5 | 09/09/79 | | Approximation Methods for Navier-Stokes Problems | F | Paderborn, FRG | 77 | 14 |
| 79-6 | 09/10/79 | | Optical Methods in Mechanics of Solids | S | Poitiers, France | 77 | 22 |
| 79-7 | 09/16/79 | | Laminar-Turbulent Transition | F | Stuttgart, FRG | 90 | 13 |
| 80-1 | 05/27/80 | | Physical Non-Linearities in Structural Analysis | S | Senlis, France | 107 | 17 |
| 80-2 | 06/02/80 | | Three-Dimensional Constitutive Relationships and Ductile Fracture | S | Dourdan, France | 96 | 14 |
| 80-3 | 08/11/80 | | Finite Elasticity | S | Bethlehem, PA, USA | 59 | 12 |
| 80-4 | 09/08/80 | | Creep in Structures, 3rd | S | Leicester, UK | 74 | 10 |
| 80-5 | 10/06/80 | ICHMT | Heat and Mass Transfer and the Structure of Turbulence | T | Dubrovnik, Yu-goslavia | 46 | 12 |
| 81-1 | 03/16/81 | IUPAC | Interaction of Particles in Colloidal Dispersions | F | Canberra, Aus-tralia | 82 | 13 |
| 81-2 | 03/23/81 | CISM | Crack Formation and Propagation | S | Tuzcno, Poland | 38 | 7 |
| 81-3 | 05/05/81 | | Unsteady Turbulent Shear Flows | F | Toulouse, France | 57 | 11 |
| 81-4 | 06/15/81 | | Mechanics and Physic of Gas Bubbles in Liquids | F | Pasadena, CA, USA | 59 | 13 |
| 81-5 | 07/14/81 | IUGG | Intense Atmospheric Vortices | G | Reading, UK | 34 | |
| 81-6 | 08/31/81 | | Stability in the Mechanics of Continua | S | Nuremberg, FRG | 59 | 18 |
| 81-7 | 09/15/81 | | High Temperature Gas Dynamics | T | Liblice, Czecho-slovakia | 36 | 9 |
| 82-1 | 03/29/82 | | Three-Dimensional Turbulent Boundary Layers | F | West Berlin, FRG | 56 | 12 |
| 82-2 | 05/18/82 | COSPAR, IAU, IAMAP, URSI | Giant Planets and Their Satellites | O | Ottawa, Canada | 53 | 6 |
| 82-3 | 05/21/82 | COSPAR, IAU, IUGS | Impact Processes of Solid Bodies | S | Ottawa, Canada | ? | |
| 82-4 | 05/25/82 | COSPAR, ESA | Fundamental Aspects of Material Sciences in Space | MA | Ottawa, Canada | 36 | 10 |
| 82-5 | 06/07/82 | ISIMM | Modern Developments in Analytical Mechanics | M | Torino, Italy | 87 | 23 |
| 82-6 | 08/16/82 | | Mechanics of Composite Materials | S | Blacksburg, VA, USA | 124 | 15 |
| 82-7 | 08/23/82 | | Nonlinear Deformation Waves | F | Tallinn, USSR | 157 | 18 |

| No. | Date | Co-Sponsor | Symposia Title | Sub-ject | Location City, Country | Partici-pants | Coun-tries |
|-----|------|-----------|----------------|----------|------------------------|---------------|------------|
| 82-8 | 08/31/82 | | Collapse -The Buckling of Structures in Theory and in Practice | S | London, UK | 70 | 18 |
| 82-9 | 08/31/82 | | Deformation and Failure of Granular Materials | S | Delft, Nether-lands | 97 | 19 |
| 82-10 | 08/31/82 | | Structure of Complex Turbulent Shear Flow | F | Marseille, France | 92 | 15 |
| 82-11 | 09/06/82 | | Metallurgical Applications of Magnetohydrodynamics | F | Cambridge, UK | 66 | 14 |
| 82-12 | 11/01/82 | | Random Vibrations and Relia-bility | D | Frankfurt/Oder, GDR | 77 | 14 |
| 83-1 | 07/04/83 | IUPAP | Mechanical Behaviour of Elec-tromagnetic Solid | S | Paris, France | 81 | 19 |
| 83-2 | 07/05/83 | | Measuring Techniques in Gas-Liquid Two-Phase Flows | F | Nancy, France | 79 | 12 |
| 83-3 | 07/13/83 | ICA | Mechanics of Hearing | B | Delft, Nether-lands | 50 | 12 |
| 83-4 | 08/29/83 | IUGG | Atmospheric Dispersion of Heavy Gases and Small Particles | E | Delft, Nether-lands | 70 | 10 |
| 83-5 | 09/05/83 | IUGG | Seabed Mechanics | S | Newcastle upon Tyne, UK | 62 | 9 |
| 83-6 | 09/05/83 | | Turbulence and Chaotic Phe-nomena in Fluids | F | Kyoto, Japan | 189 | 15 |
| 83-7 | 09/11/83 | | Mechanics of Geomaterials; Rocks, Concretes, Soils (W. Prager in Memoriam) | G | Evanston, IL, USA | 109 | 24 |
| 84-1 | 04/02/84 | | Fundamentals of Deformation and Fracture (J. Eshelby in Me-moriam) | S | Sheffield, UK | 71 | 12 |
| 84-2 | 06/19/84 | | Probabilistic Methods in the Mechanics of Solids and Struc-tures. (W. Weibull in Memori-am) | S | Stockholm, Sweden | 116 | 21 |
| 84-3 | 06/26/84 | | Influence of Polymer Additives on Velocity and Temperature Fields | T | Essen, FRG | 72 | 15 |
| 84-4 | 07/09/84 | | Laminar-Turbulent Transition, 2nd | F | Novosibirsk, USSR | 130 | 17 |
| 84-5 | 09/17/84 | | Optical Methods in the Dynam-ics of Fluids and Solids | O | Liblice Castle, CSSR | 79 | 15 |
| 85-1 | 07/01/85 | | Mechanics of Damage and Fa-tigue | S | Haifa, Tel Aviv, Israel | 64 | 10 |
| 85-2 | 07/03/85 | | Aero- and Hydroacoustics | F | Lyon, France | 150 | 14 |

| No. | Date | Co-Sponsor | Symposia Title | Subject | Location City, Country | Participants | Countries |
|-----|------|-----------|----------------|---------|-----------------------|--------------|-----------|
| 85-3 | 07/08/85 | | Hydrodynamics of Ocean Wave Utilization | F | Lisbon, Portugal | 84 | 14 |
| 85-4 | 08/05/85 | | Inelastic Behaviour of Plates and Shells | S | Rio de Janeiro, Brasil | 90 | 20 |
| 85-5 | 08/12/85 | SCJ, JSASS | Macro- and Micromechanics of High Velocity Deformation and Fracture | MN | Tokyo, Japan | 60 | 11 |
| 85-6 | 08/25/85 | | Mixing in Stratified Fluids | F | Margaret River, Australia | 60 | 13 |
| 85-7 | 09/09/85 | | Turbulent Shear-Layer/ Shock-Wave Interaction | F | Palaiseau, France | 106 | 8 |
| 85-8 | 09/16/85 | IFTOMM | Dynamics of Multibody Systems | D | Udine, Italy | 47 | 20 |
| 85-9 | 11/18/85 | | Simple and Multi-Phase Fluid Flow through Heterogeneous Permeable Materials | F | Hamilton, New Zealand | 55 | 10 |
| 86-1 | 07/19/86 | | Fluid Mechanics in the Spirit of G.I. Taylor | F | Cambridge, UK | 213 | 17 |
| 86-2 | 08/26/86 | | Boundary Layer Separation | F | London, UK | 79 | 12 |
| 86-3 | 09/01/86 | | Thermomechantcal Coupling in Solids | T | Palaiseau, France | 83 | 13 |
| 86-4 | 10/13/86 | | Interaction in Deformable Solids and Structures | S | Tokyo, Japan | 90 | 14 |
| 86-5 | 11/29/86 | | Application of Asymptotic Methods to Problems in Continuum Mechanics | M | Paris, France | 70 | 11 |
| 87-1 | 01/19/87 | | Turbulence Management and Relaminarisation | F | Bangalore, India | 67 | 8 |
| 87-2 | 04/13/87 | | Advanced Boundary Element Methods: Applications in Solid and Fluid Mechanics | M | San Antonio, TX, USA | 76 | 13 |
| 87-3 | 06/21/87 | | Nonlinear Stochastic Dynamic Engineering Systems | D | Innsbruck/Igls, Austria | 87 | 19 |
| 87-4 | 08/24/87 | ICM | Yielding, Damage and Failure of Anisotropic Solids, Antoni Sawczuk in Memoriam | S | Villard-de-Lans, France | 99 | 20 |
| 87-5 | 08/24/87 | ICA | Nonlinear Acoustics | O | Novosibirsk, USSR | 162 | 15 |
| 87-6 | 08/25/87 | | Non-Linear Water Waves | F | Tokyo, Japan | 110 | 16 |
| 87-7 | 08/31/87 | | Fundamental Aspects of Vortex Motion | F | Tokyo, Japan | 110 | 14 |
| 88-1 | 02/09/88 | | Structural Optimization | OP | Melbourne, Australia | 66 | 20 |
| 88-2 | 03/14/88 | | Recent Advances in Nonlinear Fracture Mechanics | S | Pasadena, CA, USA | 51 | 9 |

| No. | Date | Co-Sponsor | Symposia Title | Sub-ject | Location City, Country | Partici-pants | Coun tries |
|---|---|---|---|---|---|---|---|
| 88-3 | 03/21/88 | IUPAP | Elastic Waves | S | Galway, Ireland | 124 | 21 |
| 88-4 | 05/16/88 | | Liquid Metal Magnetohydrody-namics | F | Riga, USSR | 80 | 13 |
| 88-5 | 05/24/88 | | Transsonicum III | F | Göttingen, FRG | 146 | 17 |
| 88-6 | 05/30/88 | IFAC | Dynamics of Controlled Me-chanical Systems | D | Zürich, Switzer-land | 65 | 11 |
| 89-1 | 06/05/89 | IACM | Discretization Methods in Struc-tural Mechanics | S | Vienna, Austria | 81 | 15 |
| 89-2 | 07/25/89 | | Structure of Turbulence and Drag Reduction | F | Zürich, Switzer-land | 107 | 13 |
| 89-3 | 07/30/89 | | Elastic Wave Propagation and Ultrasonic Nondestructive Eval-uation | S | Boulder, CO, USA | 137 | 17 |
| 89-4 | 08/13/89 | | Topological Fluid Mechanics | F | Cambridge, UK | 119 | 15 |
| 89-5 | 08/14/89 | IAHR | Ice-Structure Interaction | G | St. John´s, New-foundland, Can-ada | 90 | 8 |
| 89-6 | 08/21/89 | | Nonlinear Dynamics in Engi-neering Systems | D | Stuttgart, FRG | 78 | 22 |
| 89-7 | 08/28/89 | | Adiabatic Waves in Liquid-Vapor Systems | F | Göttingen, FRG | 67 | 14 |
| 89-8 | 09/11/89 | | Laminar-Turbulent Transition | F | Toulouse, France | 108 | 12 |
| 90-1 | 04/02/90 | | Mechanical Modellings of New Electromagnetic Materials | D | Stockholm, Sweden | 75 | 16 |
| 90-2 | 05/23/90 | | Dynamical Problems of Rigid-Elastic Systems and Structures | D | Moscow, USSR | 48 | 12 |
| 90-3 | 05/29/90 | | Inelastic Deformation of Com-posite Materials | S | Troy, NY, USA | 70 | 13 |
| 90-4 | 06/24/90 | | Dynamics of Marine Vehicles and Structures in Waves | D | Uxbridge, UK | 86 | 21 |
| 90-5 | 07/09/90 | | Separated Flows and Jets | F | Novosibirsk, USSR | 145 | 13 |
| 90-6 | 08/20/90 | | Fluid Mechanics of Stirring and Mixing | F | La Jolla, CA, USA | 141 | 17 |
| 90-7 | 09/03/90 | | Nonlinear Hydrodynamics Sta-bility and Transition | F | Sophia-Antipolis, France | 112 | 17 |
| 90-8 | 09/04/90 | | Contact Load and Local Effects in Thin-Walled Plated and Shell Structures | S | Prague, Czecho-slovakia | 70 | 22 |
| 90-9 | 09/10/90 | | Creep in Structures | S | Cracow, Poland | 123 | 22 |
| 91-1 | 06/03/91 | | Aerothermodynamics in Combusters | T | Taipei, China/ Taipei | 129 | 8 |
| 91-2 | 06/10/91 | | Mechanical Effects of Welding | S | Luleå, Sweden | 50 | 13 |
| 91-3 | 07/01/91 | | Nonlinear Stochastic Mechanics | D | Torino, Italy | 78 | 20 |
| 91-4 | 07/01/91 | | Mechanics of Fluidized Beds | F | Stanford, CA, USA | 58 | 11 |

| No. | Date | Co-Sponsor | Symposia Title | Subject | Location City, Country | Participants | Countries |
|---|---|---|---|---|---|---|---|
| 91-5 | 07/15/91 | | Breaking Waves | F | Sydney, Australia | 63 | 15 |
| 91-6 | 07/22/91 | | Constitutive Relations for Finite Deformation of Polycrystalline Metals | S | Beijing, China | 71 | 12 |
| 91-7 | 08/19/91 | | Finite Inelastic Deformations - Theory and Application | S | Hannover, Germany | 79 | 13 |
| 91-8 | 08/26/91 | | Interpretation of Time Series from Mechanical Systems | D | Warwick, UK | 66 | 16 |
| 91-9 | 09/02/91 | | Microgravity Fluid Mechanics | F | Bremen, Germany | 82 | 13 |
| 91-10 | 10/27/91 | | Local Mechanics Concepts for Composite Material Systems | S | Blacksburg, VA, USA | 73 | 15 |
| 92-1 | 04/19/92 | | Optimal Control of Mechanical Systems | D | Moscow, Russia | 71 | 8 |
| 92-2 | 05/11/92 | | Inverse Problems in Engineering Mechanics | S | Tokyo, Japan | 97 | 11 |
| 92-3 | 08/18/92 | | Optimal Design with Advanced Materials (Niordson Anniversary) | OP | Lyngby, Denmark | 60 | 18 |
| 92-4 | 09/01/92 | | Aerothermochemistry of Spacecraft and Associated Hypersonic Flows | T | Marseille, France | 103 | 15 |
| 92-5 | 09/07/92 | | Bluff-Body Wakes, Dynamics and Instabilities | F | Göttingen, Germany | 110 | 16 |
| 92-6 | 09/13/92 | | Fluid Dynamics of High Angle of Attack | F | Tokyo, Japan | 77 | 8 |
| 92-7 | 10/12/92 | | Eddy Structure Identification in Free Turbulent Shear Flow | F | Poitiers, France | 120 | 15 |
| 93-1 | 06/07/93 | | Probabilistic Structural Mechanics: Advances in Structural Reliability Methods | S | San Antonio, TX, USA | 56 | 12 |
| 93-2 | 06/15/93 | | Computational Mechanics of Materials | S | Providence, RI, USA | 71 | 13 |
| 93-3 | 07/19/93 | | Nonlinearity and Chaos in Engineering Dynamics | D | London, UK | 78 | 23 |
| 93-4 | 07/26/93 | | Nonlinear Instability of Nonparallel Flows | F | Potsdam, NY, USA | 59 | 9 |
| 93-5 | 08/15/93 | | Nonlinear Waves and Solids | S | Victoria, British Columbia, Canada | 110 | 22 |
| 93-6 | 08/23/93 | | Identification of Mechanical Systems | S | Wuppertal, Germany | 71 | 24 |
| 93-7 | 08/31/93 | | Discrete Structural Optimization | S | Zakopane, Poland | 38 | 13 |
| 93-8 | 09/06/93 | | Bubble Dynamics and Interface Phenomena | F | Edgbaston, Birmingham, UK | 81 | 14 |

| No. | Date | Co-Sponsor | Symposia Title | Subject | Location City, Country | Participants | Countries |
|-----|------|-----------|----------------|---------|------------------------|--------------|-----------|
| 93-9 | 09/20/93 | | Fracture of Brittle Disordered Materials: Concrete, Rock and Ceramics | S | Brisbane, Australia | 87 | 17 |
| 93-10 | 10/11/93 | | Impact Dynamics | D | Beijing, China | 53 | 12 |
| 93-11 | 11/01/93 | | Numerical Simulation of Nonisothermal Flow of Visco-elastic Liquids | F | Kerkrade, Netherlands | 27 | 8 |
| 94-1 | 04/18/94 | | Liquid-Particle Interaction in Suspension Flows | F | Grenoble, France | 73 | 19 |
| 94-2 | 05/09/94 | | Waves and Liquid/Gas and Liquid/Vapor Two-Phase Systems | F | Kyoto, Japan | 63 | 10 |
| 94-3 | 08/17/94 | ISIMM | Structure and Dynamics of Nonlinear Waves in Fluids | F | Hannover, Germany | 58 | 15 |
| 94-4 | 08/23/94 | | Microstructure-Property Interactions in Composite Materials | MN | Aalborg, Denmark | 76 | 15 |
| 94-5 | 08/30/94 | | Anisotopy, Inhomogeneity and Nonlinearity in Solid Mechanics | S | Nottingham, UK | 101 | 20 |
| 94-6 | 09/05/94 | | Laminar-Turbulent Transition | F | Sendai, Japan | 82 | 10 |
| 94-7 | 09/05/94 | | Mechanical Problems in Geodynamics | G | Beijing, China | 60 | 9 |
| 94-8 | 09/05/94 | | The Active Control of Vibration | D | Bath, UK | 52 | 16 |
| 94-9 | 10/03/94 | | Size-Scale Effects in the Failure Mechanisms of Materials and Structures | S | Torino, Italy | 99 | 21 |
| 94-10 | 12/06/94 | | Mechanics and Combustion of Droplets and Sprays | T | Tainan, China/ Taipei | 72 | 9 |
| 95-1 | 03/26/95 | | Optimization of Mechanical Systems | OP | Stuttgart, Germany | 91 | 20 |
| 95-2 | 06/28/95 | | Asymptotic Methods for Turbulent Shear Flows at High Reynolds Numbers | F | Bochum, Germany | 51 | 12 |
| 95-3 | 07/03/95 | | Advances in Nonlinear Stochastic Mechanics | D | Trondheim, Norway | 69 | 16 |
| 95-4 | 07/17/95 | | Nonlinear Instability and Transition in 3D Boundary Layers | F | Manchester, UK | 81 | 13 |
| 95-5 | 07/22/95 | | Hydrodynamic Diffusion of Suspended Particles | F | Estes Park, CO, USA | 68 | 12 |
| 95-6 | 08/29/95 | | Micromechanics of Plasticity and Damage of Multiphase Materials | S | Paris, France | 83 | 18 |
| 95-7 | 09/03/95 | | Nonlinear Analysis of Fracture | S | Cambridge, UK | 77 | 15 |
| 95-8 | 09/10/95 | | Physical Limnology | E | Broome, Australia | 52 | 12 |
| 95-9 | 10/02/95 | | Combustion in Supersonic Flows | F | Poitiers, France | 60 | 10 |
| 95-10 | 10/16/95 | | Constitutive Relation in High/Very High Strain Rates | S | Noda, Japan | 47 | 12 |

| No. | Date | Co-Sponsor | Symposia Title | Sub-ject | Location City, Country | Partici-pants | Coun-tries |
|-----|------|-----------|----------------|----------|------------------------|---------------|-----------|
| 96-1 | 04/21/96 | | Interaction between Dynamics and Control in Advanced Mechanical Systems | D | Eindhoven, Netherlands | 90 | 18 |
| 96-2 | 06/30/96 | | Innovative Computational Methods for Fracture and Damage | CM | Dublin, Ireland | 103 | 19 |
| 96-3 | 07/07/96 | | Variable Density Low Speed Turbulent Flows | F | Marseille, France | 93 | 14 |
| 96-4 | 07/15/96 | | Mechanics of Granular and Porous Materials | S | Cambridge, UK | 86 | 21 |
| 97-1 | 01/07/97 | CACOFD | Lubricated Transport of Viscious Materials | S | Tobago | 33 | 9 |
| 97-2 | 03/09/97 | | Transformation Problems in Composite and Active Materials | S | Cairo, Egypt | 45 | 10 |
| 97-3 | 03/17/97 | | Non-Linear Singularities in Deformation and Flow | M | Haifa, Israel | 72 | 12 |
| 97-4 | 04/22/97 | | Variations of Domains and Free-boundary Problems in Solid Mechanics | S | Paris, France | 62 | 13 |
| 97-5 | 05/25/97 | | Simulation and Identification of Organized Structures in Flows | F | Lyngby, Denmark | 67 | 16 |
| 97-6 | 06/02/97 | IACM | Discretization Methods in Structural Mechanics II | CM | Vienna, Austria | 71 | 23 |
| 97-7 | 06/09/97 | | Material Instabilities in Solids | S | Delft, Netherlands | 51 | 18 |
| 97-8 | 07/08/97 | | Statistical Energy Analysis | O | Southampton, UK | 68 | 17 |
| 97-9 | 07/20/97 | | Rheology and Computation | CM | Sydney, Australia | 67 | 16 |
| 97-10 | 07/27/97 | | New Applications of Nonlinear and Chaotic Dynamics in Mechanics | D | Ithaca, NY, USA | 55 | 19 |
| 97-11 | 07/27/97 | | Computational Methods for Unbounded Domains | CM | Boulder, CO, USA | 37 | 12 |
| 97-12 | | | Micro- and Macrostructural Aspects of Thermoplasticity | T | Bochum, Germany | 87 | 13 |
| 97-13 | 08/31/97 | | Dynamics of Slender Vortices | F | Aachen, Germany | 46 | 8 |
| 97-14 | 09/02/97 | | Rheology of Buddies with Defects | F | Beijing, China | 31 | 10 |
| 98-1 | 05/17/98 | | Three- Dimensional Aspects of Air-Sea- Interaction | FS | Nice, France | 83 | 13 |
| 98-2 | 05/24/98 | | Synthesis in Bio Solid Mechanics | B | Copenhagen, Denmark | 48 | 14 |
| 98-3 | 06/16/98 | IUGG | Developments in Geophysical Turbulence | G | Boulder, CO, USA | 91 | 15 |

| No. | Date | Co-Sponsor | Symposia Title | Sub-ject | Location City, Country | Partici-pants | Coun tries |
|---|---|---|---|---|---|---|---|
| 98-4 | 06/21/98 | | Viscoelastic Fluid Mechanics: Effects of Molecular Modelling | F | Stanford, CA, USA | 45 | 12 |
| 98-5 | 08/03/98 | | Unilateral Multibody Dynamics | D | Munich, Germany | 38 | 12 |
| 98-6 | 08/24/98 | IFToMM | Synthesis of Nonlinear Dynamical Systems | D | Riga, Latvia | 58 | 14 |
| 98-7 | 08/31/98 | | Advanced Optical Methods and Applications in Solid Mechanics | S | Poitiers, France | 80 | 16 |
| 98-8 | 09/06/98 | IASS | Deployable Structures: Theory and Applications | S | Cambridge, UK | 83 | 18 |
| 98-9 | 09/07/98 | CISM | Mechanics of Passive & Active Flow Control (FLOWCON) | F | Göttingen, Germany | 108 | 16 |
| 99-1 | 01/04/99 | | Nonlinearity and Stochastic Structural Dynamics | D | Madras, India | 43 | 10 |
| 99-2 | 01/18/99 | | Mechanical and Electromagnetic Waves in Structured Media | S | Sydney, Australia | 32 | 11 |
| 99-3 | 05/02/99 | | Recent Developments in Non-linear Oscillations of Mechanical Systems | D | Hanoi, Vietnam | 45 | 16 |
| 99-4 | 05/30/99 | IACM/IABEM | Advanced Mathematical and Computational Mechanics Aspects of the Boundary Element Method | CM | Cracow, Poland | 60 | 17 |
| 99-5 | 06/05/99 | | Segregation in Granular Flows | F | Cape May, NJ, USA | 48 | 12 |
| 99-6 | 07/07/99 | | Nonlinear Waves in Multi-Phase Flow | F | Notre-Dame, IN, USA | 52 | 14 |
| 99-7 | 09/05/99 | | Theoretical and Numerical Methods in Continuum Mechanics of Porous Materials | M | Stuttgart, Germany | 82 | 21 |
| 99-8 | 09/13/99 | | Laminar-Turbulent Transition | F | Sedona, AZ, USA | 132 | 18 |
| 99-9 | 11/01/99 | | Geometry and Statistics of Turbulence | F | Hayama, Japan | 91 | 13 |
| 00-1 | 04/02/00 | | Creep in Structures | S | Nagoya, Japan | 81 | 15 |
| 00-2 | 06/13/00 | | Bluff Body Wakes and Vortex-Induced Vibrations | F | Marseille, France | 100 | 18 |
| 00-3 | 06/14/00 | | Mechanical Waves for Composite Structures Characterization | S | Chania, Greece | more than fifty | |
| 00-4 | 07/02/00 | | Advances in Mathematical Modelling of Atmosphere and Ocean Dynamics | G | Limerick, Ireland | 72 | 14 |
| 00-5 | 07/10/00 | | Free Surface Flows | F | Birmingham, UK | 89 | 17 |
| 00-6 | 07/16/00 | | Diffraction and Scattering in Fluid Mechanics and Elasticity | M | Manchester, UK | 58 | 12 |

| No. | Date | Co-Sponsor | Symposia Title | Subject | Location City, Country | Partici-pants | Countries |
|-----|------|-----------|----------------|---------|------------------------|---------------|-----------|
| 00-7 | 07/31/00 | | Field Analysis for Determination of Material Parameters Experimental and Numerical Aspects | MA | Kiruna, Sweden | 30 | 11 |
| 00-8 | 09/26/00 | | Smart Structures and Structronic Systems | D | Magdeburg, Germany | 78 | 16 |
| 00-9 | 12/12/00 | | Designing for Quietness | D | Bangalore, India | 51 | 8 |
| 01-1 | 03/26/01 | | Flow in Collapsible Tubes and Past Other Highly Compliant Boundaries | F | Warwick, Coventry, UK | 52 | 10 |
| 01-2 | 05/07/01 | | Material Instabilities and the Effect of Microstructure | MA | Austin, TX, USA | 65 | 14 |
| 01-3 | 06/03/01 | | Turbulent Mixing and Combustion | F | Kingston, Ontario, Canada | 84 | 11 |
| 01-4 | 06/11/01 | | Micromechanics of Martensitic Phase Transformation in Solids | S | Hong Kong, China | 39 | 12 |
| 01-5 | 06/18/01 | | Analytical and Computational Fracture Mechanics of Non homogenous Materials | M | Cardiff, UK | 79 | 16 |
| 01-6 | 08/20/01 | | Computational Mechanics of Solid Materials at Large Strains | CM | Stuttgart, Germany | 100 | 14 |
| 01-7 | 09/02/01 | | Tubes, Sheets and Singularities in Fluid Dynamics | F | Zakopane, Poland | 62 | 12 |
| 02-1 | 04/03/02 | | Micromechanics of Fluid Suspensions and Solid Composites | MN | Austin, TX, USA | 40 | 8 |
| 02-2 | 04/08/02 | | Unsteady Separated Flows | F | Toulouse, France | 140 | 12 |
| 02-3 | 05/20/02 | | Dynamics of Advanced Materials and Smart Structures | MA | Yamagata, Japan | 78 | 18 |
| 02-4 | 07/08/02 | | Asymptotics, Singularities and Homogenisation in Problems of Mechanics | M | Liverpool, UK | 72 | 12 |
| 02-5 | 08/13/02 | | Complementary-Dual-Varational Principles in Nonlinear Mechanics | M | Shanghai, China | 36 | 18 |
| 02-6 | 08/25/02 | | Nonlinear Stochastic Systems | D | Urbana-Champaign, IL, USA | 76 | 13 |
| 02-7 | 09/02/02 | | Transsonicum IV | F | Göttingen, Germany | 64 | 13 |
| 02-8 | 09/11/02 | | Reynolds Number Scaling in Turbulent Flow | F | Princeton, NJ, USA | 89 | 13 |
| 02-9 | 09/24/02 | | Evolutionary Methods in Mechanics | M | Cracow, Poland | 50 | 18 |
| 02-10 | 10/20/02 | | Multiscale Modeling and Characteriziation of Elastic-Inelastic Behaviour of Engineering Materials | S | Marrakesch, Morocco | 75 | 17 |

| No. | Date | Co-Sponsor | Symposia Title | Sub-ject | Location City, Country | Partici-pants | Coun tries |
|---|---|---|---|---|---|---|---|
| 03-1 | 05/18/03 | | Mechanics of Physiochemical and Electromechanical Interactions in Porous Media | MA | Kerkrade, Netherlands | 60 | 15 |
| 03-2 | 06/02/03 | | Integrated Modeling of Fully Coupled Fluid-Structure Interactions | FS | Rutgers, NJ, USA | 55 | 12 |
| 03-3 | 06/08/03 | | Chaotic Dynamics and Control of Systems and Processes in Mechanics | D | Rome, Italy | 63 | 20 |
| 03-4 | 07/06/03 | | Mesoscopic Dynamics of Fracture Process and Materials Strength | S | Osaka, Japan | 66 | 8 |
| 04-1 | 05/30/04 | | Size Effects on Material and Structural Behavior at Micron- and Nanometer-Scales | MN | Hong Kong, China | 60 | 8 |
| 04-2 | 06/27/04 | | Mechanics of Biological Tissue | B | Graz, Austria | 96 | 19 |
| 04-3 | 08/09/04 | | Size Effects on Non-Uniqueness of Solutions to the Navier-Stokes Equations and their Connection with Laminar-Turbulent Transition | F | Manchester, UK | 32 | 9 |
| 04-4 | 08/12/04 | | One Hundred Years of Boundary Layer Research | F | Göttingen, Germany | 77 | 16 |
| 04-5 | 09/01/04 | | Elastohydrodynamics and Microelastohydrodynamics | MN | Cardiff, UK | 45 | 10 |
| 04-6 | 09/01/04 | | Mechanics and Reliability of Actuating Materials | MA | Beijing, China | 55 | 8 |
| 04-7 | 10/04/04 | | Recent Advances in Disperse Multiphase Flow Simulation | F | Urbana-Champaign, IL, USA | 90 | 16 |
| 04-8 | 10/26/04 | | Elementary Vortices and Coherent Structures: Significance in Turbulence Dynamics | F | Kyoto, Japan | 79 | 11 |
| 04-9 | 12/13/04 | | Laminar-Turbulent Transition | F | Bangalore, India | 113 | 15 |
| 05-1 | 05/23/05 | | Multiscale Modelling of Damage and Fracture Processes in Composite Materials | S | Kazimierz Dolny, Poland | 48 | 15 |
| 05-2 | 06/27/05 | | Mechanical Behavior and Micro-mechanics of Nanostructured Materials | MN | Beijing, China | 33 | 10 |
| 05-3 | 07/11/05 | | Impact Biomechanics - From Fundamental Insights to Applications | B | Dublin, Ireland | 94 | 19 |
| 05-4 | 07/18/05 | | Vibration Control of Nonlinear Mechanisms and Structures | D | Munich, Germany | 68 | 14 |

| No. | Date | Co-Sponsor | Symposia Title | Sub-ject | Location City, Country | Partici-pants | Coun tries |
|---|---|---|---|---|---|---|---|
| 05-5 | 10/26/05 | | Topological Design Optimiza-tion of Structures, Machines and Materials-Status and Perspec-tives | OP | Rungstegaard, Denmark | 72 | 19 |
| 06-1 | 02/20/06 | | Multiscale Problems in Multibody System Contacts | D | Stuttgart, Germany | 57 | 17 |
| 06-2 | 03/26/06 | | Interactions for Dispersed Sys-tems in Newtonian and Viscoe-lastic Fluids | F | Guanajuato, Mexico | 48 | 13 |
| 06-3 | 05/21/06 | | Plasticity at the Micron Scale | MN | Lyngby, Denmark | 66 | 13 |
| 06-4 | 08/25/06 | | Hamiltonian Dynamics, Vortex Structures, Turbulence | M | Moscow, Russia | 73 | 11 |
| 06-5 | 09/04/06 | | Discretization Methods for Evolving Discontinuities | CM | France, Lyon | 38 | 10 |
| 06-6 | 09/11/06 | | Computational Physics and New Perspectives in Turbulence | CM | Japan, Nagoya | 104 | 13 |
| 06-7 | 09/18/06 | | Dynamics and Control of Non-linear Systems with Uncertainty | D | Nanjing, China | 41 | 14 |
| 06-8 | 09/19/06 | | Flow Control with Mems | F | London, UK | 120 | 14 |
| 06-9 | 11/05/06 | | Computational Contact Mechan-ics | CM | Hannover, Germany | 39 | 11 |
| 07-1 | 04/23/07 | | Relation of Shell, Plate, Beam and 3D Models | S | Tbilisi, GA, USA | 50 | 15 |
| 07-2 | 06/11/07 | | Recent Advances in Multiphase Flows: Numerical and Experi-mental | F | Istanbul, Turkey | 56 | 14 |
| 07-3 | 06/18/07 | | Unsteady Separated Flows and their Control | F | Corfu, Greece | 140 | 12 |
| 07-4 | 06/25/07 | | Scaling in Solid Mechanics | S | Cardiff, UK | 77 | 20 |
| 07-5 | 07/23/07 | | Fluid-Structure Interaction in Ocean Engineering | G | Hamburg, Ger-many | 40 | 15 |
| 07-6 | 08/06/07 | | Swelling and Shrinking of Po-rous Materials: From Colloid Science to Pro-Mechanics | O | Petropolis, Brazil | 70 | 15 |
| 07-7 | 09/06/07 | | Advances in Micro- and Nanofluidics | MN | Dresden, Ger-many | 28 | 12 |
| 07-8 | 09/17/07 | | Mechanical Properties of Cellu-lar Materials | MA | Cachan, France | 40 | 14 |
| 07-9 | 11/05/07 | | Multi-Scale Plasticity of Crys-talline Materials | MA | Eindhoven, Netherlands | 76 | 13 |
| 08-1 | 01/14/08 | | Theoretical, Modelling and Computational Aspects of Ine-lastic Media | S | Cape Town, South Africa | 59 | 15 |
| 08-2 | 05/19/08 | | Modelling Nanomaterials and Nanosystems | MN | Aalborg, Denmark | 53 | 19 |

| No. | Date | Co-Sponsor | Symposia Title | Subject | Location City, Country | Participants | Countries |
|---|---|---|---|---|---|---|---|
| 08-3 | 06/18/08 | | Cellular, Molecular and Tissue Mechanics | MA | Cape Cod, Wood Hole, MA, USA | 62 | 9 |
| 08-4 | 09/22/08 | | Variational Concepts with Applications to the Mechanics of Materials | MA | Bochum, Germany | 56 | 10 |
| 08-5 | 10/12/08 | | 150 Years of Vortex Dynamics | F | Lyngby and Copenhagen, Denmark | 108 | 20 |
| 08-6 | 10/20/08 | | Progress in the Theory and Numerics of Configurational Mechanics | M | Erlangen, Germany | 49 | 10 |
| 08-7 | 12/08/08 | | Rotating Stratified Turbulence and Turbulence in the Atmosphere and Oceans | G | Cambridge, UK | 86 | |
| 08-8 | 12/10/08 | | Multi Functional Material Structures and Systems | MA | Bangalore, India | 74 | 10 |
| 09-1 | 03/08/09 | | Dynamic Fracture and Fragmentation | S | Austin, TX, USA | 55 | 5 |
| 09-2 | 03/23/09 | | Emerging Trends in Rotor Dynamics | D | New Delhi, India | 86 | 21 |
| 09-3 | 05/25/09 | | Recent Advances of Acoustic Waves in Solids | S | Taipei, China/Taipei | 50 | 16 |
| 09-4 | 06/23/09 | | Laminar-Turbulent Transition | F | Stockholm, Sweden | 142 | 21 |
| 09-5 | 06/05/09 | | Symposium on Vibration Analysis of Structures with Uncertainties | D | St. Petersburg, Russia | 37 | 13 |
| 09-6 | 07/07/09 | | Wall-bounded Turbulent Flows on Rough Walls | F | Cambridge, UK | 44 | 11 |
| 09-7 | 09/01/09 | | Multiscale Modelling of Fatigue, Damage and Fracture in Smart Materials Systems | MA | Freiberg, Germany | 44 | 14 |
| 09-8 | 09/14/09 | ISIMM | Mathematical Modeling and Physical Instances of Granular Flows | F | Reggio Calabria, Italy | 56 | 15 |
| 10-1 | 03/29/10 | | Computational Aero-Acoustics for Aircraft Noise Prediction | CM | Southampton, UK | 58 | 10 |
| 10-2 | 05/10/10 | | Nonlinear Stochastic Dynamics and Control | D | Hangzhou, China | 44 | 13 |
| 10-3 | 06/07/10 | | Dynamics Modeling and Interaction Control in Virtual and Real Environments | D | Budapest, Hungary | 45 | 17 |
| 10-4 | 06/22/10 | | Bluff Body Wakes and Vortex-Induced Vibrations | F | Capri, Italy | 107 | 22 |
| 10-5 | 07/27/10 | | Nonlinear Dynamics for Advanced Technologies and Engineering Design (NDATED) | D | Aberdeen, UK | 73 | 19 |

| No. | Date | Co-Sponsor | Symposia Title | Sub-ject | Location City, Country | Partici-pants | Coun-tries |
|---|---|---|---|---|---|---|---|
| 10-6 | 08/08/10 | | Surface Effects in the Mechanics of Nanomaterials and Heterostructures | MN | Beijing, China | 40 | 9 |
| 10-7 | 09/13/10 | | Analysis and Simulation of Human Motion | B | Leuven, Belgium | 106 | 15 |
| 11-1 | 05/08/11 | | Mechanics of Liquid and Solid Foams | F | Austin, TX, USA | 50 | 14 |
| 11-2 | 05/17/11 | | Linking Scales in Computations: From Microstructure to Macroscale Properties | MN | Pensacola, FL, USA | 33 | 7 |
| 11-3 | 06/05/11 | | Human Body Dynamics: From Multibody Systems to Biomechanics | B | Waterloo, Canada | 35 | 11 |
| 11-4 | 07/04/11 | | Full-field Measurements and Identification in Solid Mechanics | S | Cachan, France | 45 | 10 |
| 11-5 | 07/07/11 | | Impact Biomechanics in Sport | B | Dublin, Ireland | 55 | 13 |
| 11-6 | 08/29/11 | | Computer Models in Biomechanics: from Nano to Macro | B | Stanford, CA, USA | 68 | 14 |
| 11-7 | 11/28/11 | | 50 Years of Chaos: Applied and Theoretical | M | Kyoto, Japan | 185 | 15 |
| 11-8 | 12/12/11 | | Bluff Body Flows | F | Kanpur, India | 108 | 15 |
| 12-1 | 01/23/12 | | Mobile Particulate Systems – Kinematics, Rheology and Complex Phenomena | M | Bangalore, India | 72 | 12 |
| 12-2 | 04/23/12 | | Advanced Materials Modelling for Structures | MA | Paris, France | 61 | 11 |
| 12-3 | 06/05/12 | | From Mechanical to Biological Systems - an Integrated Approach | B | Izhevsk, Russia | 63 | 8 |
| 12-4 | 06/18/12 | | Waves in Fluids: Effects of Nonlinearity, Rotation, Stratification and Dissipation | F | Moscow, Russia | 171 | 12 |
| 12-5 | 06/25/12 | | Multiscale Problems in Stochastic Mechanics | D | Karlsruhe, Germany | 35 | 9 |
| 12-6 | 07/01/12 | | Fracture Phenomena in Nature and Technology | S | Brescia, Italy | 65 | 13 |
| 12-7 | 07/02/12 | | Understanding Common Aspects of Extreme Events in Fluids | F | Dublin, Ireland | 57 | 12 |
| 12-8 | 07/23/12 | | Topological Fluid Dynamics II | F | Cambridge, UK | 79 | 14 |
| 12-9 | 08/15/12 | | Hysteresis and Pattern Evolution in Non-Equilibrium Solid-Solid Phase Transitions | S | Hong Kong, China | 50 | 8 |
| 12-10 | 10/15/12 | | Particle Methods in Fluid Mechanics | M | Lyngby, Denmark | 26 | 14 |

Appendix 13: Symposia by Subject, Location, and Number of Participants

| No. | Date | Co-Sponsor | Symposia Title | Sub-ject | Location City, Country | Partici-pants | Coun-tries |
|---|---|---|---|---|---|---|---|
| 12-11 | 11/03/12 | | Advances of Optical Methods in Experimental Mechanics | O | Taipei, China/ Taipei | 53 | 7 |
| 13-1 | 03/10/13 | | Vortex Dynamics: Formation, Structure and Function | F | Fukuoka, Japan | 133 | 16 |
| 13-2 | 04/14/13 | | Nonlinear Interfacial Wave Phe-nomena from the Micro- to the Macro-scale | MN | Limassol, Cyprus | 45 | 10 |
| 13-3 | 05/06/13 | | Recent Development of Experi-mental Techniques Under Im-pact Loading | S | Xi 'an, China | 45 | 9 |
| 13-4 | 06/17/13 | | Materials and Interfaces under High Strain Rate and Large De-formation | S | Metz, France | 63 | 13 |
| 13-5 | 09/09/13 | | Multiscale Modeling and Uncer-tainty Quantification of Materi-als and Structures | S | Santorini Island, Greece | 39 | 12 |
| 13-6 | 09/23/13 | | The Dynamics of Extreme Events Influenced by Climate Change | G | Lanzhou, China | 51 | 10 |
| 14-1 | 01/20/14 | | Transition and Turbulence in the Flow through Deformable Tubes and Channels | F | Bangalore, India | 45 | 4 |
| 14-2 | 05/12/14 | | Mechanics of Soft Active Mate-rials | MA | Haifa, Israel | 43 | 12 |
| 14-3 | 05/13/14 | | Connecting Multiscale Me-chanics to Complex Material Design | MA | Evanston, IL, USA | 46 | 10 |
| 14-4 | 06/09/14 | | Micromechanics of Defects in Solids | S | Sevilla, Spain | 50 | 11 |
| 14-5 | 06/09/14 | | Dynamical Analysis of Multibody Systems with De-sign Uncertainties | D | Stuttgart, Germany | 23 | 14 |
| 14-6 | 06/16/14 | | Thermomechanical-electromagnetic Coupling in Solids: Microstructural and Stability Aspects | S | Paris, France | 30 | 4 |
| 14-7 | 07/15/14 | | Dynamics of Capsules, Vesicles and Cells in Flow | F | Compiègne, France | 60 | 13 |
| 14-8 | 09/01/14 | | Innovative Numerical Ap-proaches for Materials and Structures in Multi-field and Multi-scale Problems | CM | Attendorn, Germany | 52 | 9 |
| 14-9 | 09/08/14 | | Complexity of Nonlinear Waves | F | Tallinn, Estonia | 59 | 18 |

| No. | Date | Co-Sponsor | Symposia Title | Sub-ject | Location City, Country | Partici-pants | Coun-tries |
|---|---|---|---|---|---|---|---|
| 14-10 | 12/08/14 | | Multiphase Flows with Phase Change: Challenges and Op-portunities | F | Hyderabad, India | 140 | 10 |
| 14-11 | 12/15/14 | | Computation, Modeling and Control of Transitional and Turbulent Flows | F | Goa, India | 137 | 18 |
| 15-1 | 3/9/15 | | Dynamics of Bubbly Flows | F | Oaxaca, Mexico | 43 | |
| 15-2 | 3/17/15 | | Plastic Localization and Ductile Failure | MA | Paris, France | 50 | 11 |
| 15-3 | 6/23/15 | | Growing Solids | S | Moscow, Russia | 88 | 12 |
| 15-4 | 7/6/15 | | Analytical Methods in Nonlinear Dynamics | D | Frankfurt, Germany | 68 | 20 |
| 15-5 | 12/9/15 | | Multiphase Continuum Model-ling of Particulate Flows | F | Gainesville, USA | 25 | 6 |

Subject codes used in classifying symposia:

| | |
|---|---|
| M | Foundations and Basic Methods in Mechanics |
| D | Vibration, Dynamics, Control |
| S | Mechanics of Solids |
| F | Mechanics of Fluids |
| T | Thermal Sciences |
| G | Earth Sciences |
| E | Energy Systems and Environment |
| B | Biosciences |
| O | Other |
| FS | Fluid Structure Interaction |
| MA | Materials |
| MN | Micro-Nanomechanics |
| CM | Computational Methods |
| OP | Optimization |

# Appendix 14
# Symposia Publications

| Symposia | Total No. Papers | Symposia Name | Editors of Books (or Special Issue of Journals) | Publisher | City | Year | Name | Volume No. | Issue No. |
|---|---|---|---|---|---|---|---|---|---|
| 49-1 | | Problems on Motion of Gaseous Masses of Cosmical Dimensions, 1st | | Cent. Air Documents Off. | Dayton, OH, USA | 1951 | | | |
| 50-1 | | Colloque de Mécanique Pallanza | | | | 1950 | La Ricerca Scientifica | 20 | 12 |
| 50-2 | | Plastic Flow and Deformation within the Earth | | | | 1951 | Transactions American Geophysical Union | 32 | 4 |
| 51-1 | 19 | Non·Linear Vibrations | | El vente au Service de Docmentation et d'Information Tech Technique de l'aeronautique | Paris, France | 1953 | Publications Sci., Min. de l'Air | 281 | |
| 52-1 | 24 | Colloques sur les Zones Arides | | Palais des Academies | Bruxelles, Belgium | 1954 | Academie Royale des Sciences de Belgique. Classe des Sciences. Memoires. Collection in octavo. 2 ser. | 28 | 6 |
| 53-1 | 31 | Gas Dynamics of the Interstellar Cloud, 2nd | Van de Hulst, H.C. and Burgers, J.M. | North-Holland Pub. Co. | Amsterdam, Netherlands | 1955 | | | |
| 54-1 | 36 | Photoélasticité et Photoplasticité | | Imprimerie Dionere | Brussels, Belgium | 1954 | | | |
| 55-1 | 35 | Colloquium on Fatigue | Weibull, W. and Odqvist, F.K.G. | Springer-Verlag | Berlin, Germany | 1956 | | | |
| 55-2 | 33 | Deformalion and Flow of Solids | Grammel, R. | Springer-Verlag | Berlin, Germany | 1956 | | | |
| 57-1 | 42 | Cosmical Gas Dynamics, 3rd | Burgers, J.M. and Thomas, R.N. | Am. Inst. of Physics | New York, NY, USA | 1958 | Reviews of Modern Physics | 30 | 3 |
| 57-2 | 30 | Couche-Limite | Görtler, H. | Springer-Verlag | Berlin, Germany | 1958 | | | |

© The Author(s) 2016
P. Eberhard and S. Juhasz (eds.), *IUTAM*,
DOI 10.1007/978-3-319-31063-3

| Symposia | Total No. Papers | Symposia Name | Editors of Books (or Special Issue of Journals) | Publisher | City | Year | Name | Volume No. | Issue No. |
|---|---|---|---|---|---|---|---|---|---|
| 58-1 | 43 | Atmospheric Diffusion and Air Pollution | Frenkiel, F.N. and Sheppard, P.A. | Academic Press | New York, USA | 1959 | Advances in Geophysics | 6 | |
| 58-2 | 55 | Non-Homogénéité en Elasticité et Plasticité | Olszak, W. | Pergamom Press | London, UK | 1959 | | | |
| 59-1 | 12 | Fluid Mechanics in the Ionosphere | Bolgiano, R. | | | | 1959 | Journal of Geophysical Research | 64 | 12 |
| 59-2 | 23 | Theory of Thin Elastic Shells | Koiter, W.T. | Technological University and Interscience | Delft, Netherlands New York | 1960 | | | |
| 60-1 | 51 | Magneto-Fluid Dynamics | Frenkiel, F.N. and Sears, W.R. | NAS/NRC | Washington, D.C., USA | 1960 | Reviews of Modern Physics | 32 | 4 |
| 60-2 | 20 | Creep in Structures | Hoff, N.J. | Springer-Verlag | Berlin, Germany | 1962 | | | |
| 60-3 | 9 | Cosmical Gas Dynamics, 4th | Thomas, R.N. | N. Zanichelli | Bologna, Italy | 1961 | Supplemento del Nuovo Cimento | 22 | 1 |
| 61-1 | 41 | Fundamental Problems in Turbulence and Their Relation to Geophysics | Frenkiel, F.N. | American Geophysical Union | Washington, D.C. USA | 1962 | Journal of Geophysical Research | 67 | 8 |
| 61-2 | ? | Theory of Non-Linear Vibrations | | Academy of Sciences of the USSR | Kiev, USSR | 1963 | | | |
| 62-1 | ? | Second-Order Effects in Elasticity, Plasticity and Fluid Dynamics | Reiner, M. and Abir, D. | Macmillan | New York, NY, USA | 1964 | | | |
| 62-2 | 25 | La Dynamique des Satellites and Dynamics of Satellites | Roy, M. | Springer-Verlag and Academic Press | Berlin, Germany and New York, NY, USA | 1963 | | | |
| 62-3 | 23 | Gyrodynamics | Ziegler, H. | Springer-Verlag | Berlin, Germany | 1963 | | | |
| 62-4 | 33 | Transsonicum, 1st | Oswatitsch, K. | Springer-Verlag | Berlin, Germany | 1964 | | | |
| 63-1 | 23 | Stress Waves in Anelastic Solids | Kolsky, H. and Prager, W. | Springer-Verlag | Berlin, Germany | 1964 | | | |

| Symposia | Total No. Papers | Symposia Name | Editors of Books (or Special Issue of Journals) | Publisher | City | Year | Name | Volume No. | Issue No. |
|---|---|---|---|---|---|---|---|---|---|
| 63-2 | 59 | Applications of the Theory of Functions in Continuum Mechanics | Muskhelishvili, N.I., Sedov, L.I., Mikhailov, G.K. | Nauka Publishing House | Moscow, USSR | 1965 | | | |
| 64-1 | 38 | Rhéologie et Mécanique des Sols | Kravtchenko, J. and Sirieys, P.M. | Springer-Verlag | Berlin, Germany | 1966 | | | |
| 64-2 | 69 | Concentrated Vortex Motions in Fluids | Kuchemann, D. | | | 1965 | Journal of Fluid Mechanics | 21 | 1 |
| 65-1 | | Recent Advance in Linear Vibration Mechanics | | Societe Francaise des Mecaniciens | | 1965 | Revue Francaise de Mecanique | | 13-15 |
| 65-2 | 26 | Trajectories of Artificial Celestial Bodies as Determined from Observation | Kovalevsky, J. | Springer-Verlag | Berlin, Germany | 1996 | | | |
| 65-3 | 10 | Cosmical Gas Dynamics, 5th | Thomas, R.N. | Academic Press | London and New York | 1967 | | | |
| 66-1 | 45 | Rotating Fluid Systems | Bretherton, F.P., Carrier, G.F., and Longuet Higgins, M.S. | | | 1966 | Journal of Fluid Mechanics | 26 | 2 |
| 66-2 | 23 | Irreversible Aspects of Continuum Mechanics | Parkus, H. and Sedov, L.I. | Springer-Verlag | | 1968 | | | |
| 66-3 | 7 | Transfer of Physical Characteristics in Moving Fluids | | Springer-Verlag | Vienna and New York | 1968 | | | |
| 66-4 | 79 | Boundary Layers and Turbulence Including Geophysical Applications | Bowden, K.F., Frenkiel, F.N. and Tani, I. | Am. Institute of Physics | New York, NY, USA | 1967 | Physics of Fluids | 10 | 9, Pt.2 |
| 67-1 | 44 | Generalized Cosserat Continuum and Continuum Theory of Dislocations with Application | Kröner, E. | Springer-Verlag | Berlin, Germany | 1968 | | | |
| 67-2 | 23 | Theory of Thin Shells, 2nd | Niordson, F.I. | Springer-Verlag | Berlin, Germany | 1969 | | | |
| 67-3 | 60 | Behaviour of Dense Media Under High Dynamic Pressures | Berger, J. | Dunod and Gordon and Breach | Paris, France and New York | 1968 | | | |

| Symposia | Total No. Papers | Symposia Name | Editors of Books (or Special Issue of Journals) | Publisher | City | Year | Name | Volume No. | Issue No. |
|---|---|---|---|---|---|---|---|---|---|
| 68-1 | 18 | Thermo-inelasticity | Boley, B.A. | Springer | Vienna and New York | 1970 | | | |
| 68-2 | 56 | High-Speed Computing in Fluid Dynamics | Frenkiel, F.N. and Stewartson, K. | Am. Institute of Physics | New York, NY, USA | 1969 | Physics of Fluids, Supplement II | 12 | |
| 69-1 | 54 | Flow of Fluid-Solid Mixtures | Davidson, J.F. and Pearson, J.R.A. | Cambridge Univ. Press | Cambridge, USA | 1969 | Journal of Fluid Mechanics | 39 | 2 |
| 69-2 | 44 | Electro-hydrodynamics | Melcher, J.R. | Cambridge Univ. Press | Cambridge, USA | 1970 | Journal of Fluid Mechanics | 40 | 3 |
| 69-3 | 47 | Dynamics of Satellites | Morandu, M.B. | Springer-Verlag | Berlin, Germany | 1970 | | | |
| 69-4 | 57 | Instability of Continuous Systems | Leipholz, H. | Springer-Verlag | Berlin, Germany | 1971 | | | |
| 69-5 | ? | Cosmical Gas Dynamics, 6th | Habing, H.J. | D. Reidel Publishing Company | Dordrecht, Holland | 1970 | | | |
| 70-1 | 33 | Creep in Structures, 2nd | Hult, J. | Springer-Verlag | Berlin, Germany | 1972 | | | |
| 70-2 | 31 | High-Speed Computing of Elastic Structures | Fraeijs de Veubeke, B. | Universite de Liege | Liege, Belgium | 1971 | | | |
| 71-1 | 27 | Flow of Multiphase Fluids in Media | | | Calgary, Canada | 1971 | | | |
| 71-2 | 58 | Unsteady Boundary Layers | Eichelbrenner, E.A. | Les Presses de Universite Laval | Quebec, Canada | 1972 | | | |
| 71-3 | 41 | Non-steady Flow of Water at High Speeds | Sedov, L.I., Stepanov, G.Yu. | Nauka Publishing House | Moscow, USSR | 1973 | | | |
| 71-4 | 38 | Dynamics of Ionized Gases | Lighthill, M.J., Imai, I. and Sato, H. | University of Tokyo Press and J. Wiley | Tokyo and New York | 1973 | | | |
| 72-1 | 29 | Directional Stability and Control of Bodies Moving in Water | Bishop, R.E.D., Parkinson, A.G. and Taylor, R.E. | Institution of Mechanical Engineers | London | 1973 | Journal of Mechanical Engineering Science, Suppl. Issue 1972 | 14 | 7 |
| 72-2 | 25 | Stability of Stochastic Dynamical Systems | Curtain, R.F. | Springer-Verlag | Berlin, Germany | 1972 | Lecture Notes in Mathematics | 294 | |
| 72-3 | 50 | Flow-Induced Structural Vibrations | Naudascher, E. | Springer-Verlag | Berlin, Germany | 1974 | | | |

| Symposia | Total No. Papers | Symposia Name | Editors of Books (or Special Issue of Journals) | Publisher | City | Year | Name | Volume No. | Issue No. |
|---|---|---|---|---|---|---|---|---|---|
| 73-1 | 61 | Turbulent Diffusion in Environmental Pollution | Frenkiel, EN. and Munn, R.E. | Academic Press | New York, NY, USA | 1974 | Advances in Geophysics | 18 | |
| 73-2 | 41 | Optimization in Structural Design | Sawezuk, A. and Mroz, Z. | Springer-Verlag | Berlin, Germany | 1975 | | | |
| 73-3 | 41 | Stability of the Solar System and of Small Stellar Systems | Kozai, Y. | D. Rudel Publishing Co. | Dordrecht, Holland | 1974 | | | |
| 73-4 | 27 | Photoelastic Effect and its Applications | Kestens, J. | Springer-Verlag | Berlin, Germany | 1975 | | | |
| 74-1 | 29 | Buckling of Structures | Budiansky, B. | Springer-Verlag | Berlin, Germany | 1976 | | | |
| 74-2 | 23 | Satellite Dynamics | Giacaglia, G.E.O. | Springer-Verlag | Berlin, Germany | 1975 | | | |
| 74-3 | 22 | Dynamics of Rotors | Niordson, F.I. | Springer-Verlag | Berlin, Germany | 1975 | | | |
| 74-4 | 26 | Mechanics of the Contact Between Deformable Bodies | Pater, A.D. de and Kalker, J.J. | Delft University Press | Delft, Netherlands | 1975 | | | |
| 74-5 | 27 | Mechanics of Visco-Elastic Media and Bodies | Hult, J. | Springer-Verlag | Berlin, Germany | 1975 | | | |
| 75-1 | 52 | Dynamics of Vehicles on Roads and on Railway Tracks | Pacejka, H.B. | Swets and Zeitlinger | Amsterdam, Netherlands | 1976 | | | |
| 75-2 | 31 | Biodynamics of Animal Locomotion | Pedley, T.J. | Academic Press | London, UK | 1977 | | | |
| 75-3 | 43 | Applications or Methods of Functional Analysis to Problems of Mechanics | Germain, P. and Nayroles, B. | Springer-Verlag | Berlin, Germany | 1976 | Lecture Notes in Mathematics | 503 | |
| 75-4 | 61 | Transsonicum, 2nd | Oswatitsch, K. and Rues, D. | Springer-Verlag | Berlin, Germany | 1976 | | | |
| 76-1 | 45 | Structure of Turbulence and Drag Reduction | Frenkiel, F.N., Landahl, M.T. and Lumley, J.L. | Am. Inst. of Physics | New York, NY | 1977 | The Physics of Fluids | 20 | 10, pt. 2 |
| 76-2 | 28 | Stochastic Problems in Dynamics | Clarkson, B. L. | Pitman Publishing Ltd. | London, UK | 1977 | | | |
| 76-3 | 23 | Surface Gravity Waves in Water of Varying Depth | Provis, D.G. and Radok, R. | Springer-Verlag | Berlin, Germany | 1977 | Lecture Notes in Physics | 64 | |

| Symposia | Total No. Papers | Symposia Name | Editors of Books (or Special Issue of Journals) | Publisher | City | Year | Name | Volume No. | Issue No. |
|---|---|---|---|---|---|---|---|---|---|
| 76-4 | 24 | Aeroelasticity in Turbomachines | | | | 1976 | Revue Francaise de Mecanique, Special Issue | | |
| 77-1 | 39 | High Velocity Deformation of solids | Kawata, K. and Shioiri, J. | Springer-Verlag | Berlin, Germany | 1978 | | | |
| 77-2 | 30 | Dynamics of Multibody Systems | Magnus, K. | Springer-Verlag | Berlin, Germany | 1978 | | | |
| 77-3 | 42 | Modern Problems in Elastic Wave Propagation | Miklowitz, J. and Achenbach, J.D. | John Wiley & Sons | New York, NY, USA | 1978 | | | |
| 77-4 | 42 | Dynamics of Vehicles on Roads and Tracks | Slibar, A. and Springer, H. | Swets & Zeitlinger | Amsterdam, Netherlands | 1978 | | | |
| 77-5 | 51 | Monsoon Dynamics | Lighthill, J. and Pearce, R.P. | Cambridge University Press | Cambridge, UK | 1981 | | | |
| 78-1 | | no proceedings | | | | | | | |
| 78-2 | 39 | Shell Theory | Koiter, W.T. and Mikhailov, G.K. | North-Holland Publishing Co. | Amsterdam, Netherlands | 1980 | | | |
| 78-3 | 23 | Group Theoretical Methods in Mechanics | Ibragimov, N.H. and Orsjannikov, L.V. | USSR Academy of Sciences | Siberia, USSR | 1978 | | | |
| 78-4 | 28 | Non-Newtonian Fluid Mechanics | Walters, K. and Crochet, M.J. | | | 1979 | Journal of Non-Newtonian Fluid Mechanics | 5 | |
| 78-5 | 25 | Metal Forming Plasticity | Lippmann, H. | Springer-Verlag | Berlin, Germany | 1979 | | | |
| 78-6 | 64 | Variational Methods in the Mechanics of Solids | Nemat-Nasser, S. | Pergamom Press | Oxford, UK | 1980 | | | |
| 79-1 | 37 | Structural Control | Leipholz, H.H.E. | North-Holland Publishing Co. | Amsterdam, Netherlands | 1980 | | | |
| 79-2 | 26 | Physics and Mechanics of Ice | Tryde, P. | Springer-Verlag | Berlin, Germany | 1980 | | | |
| 79-3 | 40 | Mechanics of Sound Generation in Flows | Müller, E.A. | Springer-Verlag | Berlin, Germany | 1979 | | | |

| Symposia | Total No. Papers | Symposia Name | Editors of Books (or Special Issue of Journals) | Publisher | City | Year | Name | Volume No. | Issue No. |
|---|---|---|---|---|---|---|---|---|---|
| 79-4 | 71 | Practical Experiences with Flow Induced Vibralions | Naudascher, E. and Rockwell, D. | Springer-Verlag | Berlin, Germany | 1980 | | | |
| 79-5 | 36 | Approximation Methods for Navier-Stokes Problems | Rautmann, R. | Springer-Verlag | Berlin, Germany | 1980 | | | |
| 79-6 | 55 | Optical Methods in Mechanics of Solids | Lagarde, A. | Sijthoff & Noordhoff | Alphen aan den Rijn, Netherlands | 1981 | | | |
| 79-7 | 34 | Laminar-Turbulent Transition | Eppler, R. | Springer-Verlag | Berlin, Germany | 1980 | | | |
| 80-1 | 47 | Physical Non-linearities in Structural Analysis | Hull, J. and Lemaitre, J. | Springer-Verlag | Berlin, Germany | 1981 | | | |
| 80-2 | 39 | Three-Dimensional Constitutive Relations and Ductile Fracture | Nemat-Nasser, S. | North-Holland Publishing Co. | Amsterdam, Netherlands | 1981 | | | |
| 80-3 | 29 | Finite Elasticity | Carlson, D.E. and Shield, R.T. | Kluwer Boston and M. Nijhoff | Boston and The Hague | 1982 | Journal of Elasticity | | |
| 80-4 | ? | Creep in Structures, 3rd | Ponter, A.R.S. and Hayhurst, D.R. | Springer-Verlag | Berlin, Germany | 1981 | | | |
| 80-5 | 32 | Heat and Mass Transfer and the Structure of Turbulence | Zaric, Z.P. | Hemisphere Publishing | Washington | 1982 | | | |
| 81-1 | 52 | Interaction of Particles in Colloidal Dispersions | | | | | Advances in Colloid and Interface Science, Special Issue | | |
| 81-2 | | no proceedings | | | | | | | |
| 81-3 | 34 | Unsteady Turbulent Shear Flow | Michel, R., Cousteix, J., Houdeville, R. | Springer-Verlag | Berlin, Germany | 1981 | | | |
| 81-4 | 36 | Mechanics and Physics of Gas Bubbles in Liquids | van Wijngaarden, L. | Martinus Nijhoff Publishers | Hague, Netherlands | 1982 | | | |
| 81-5 | 26 | Intense Atmospheric Vortices | Bengtsson, L. and Lighthill, J. | Springer-Verlag | Berlin, Germany | 1982 | | | |

| Symposia | Total No. Papers | Symposia Name | Editors of Books (or Special Issue of Journals) | Publisher | City | Year | Name | Volume No. | Issue No. |
|---|---|---|---|---|---|---|---|---|---|
| 81-6 | 37 | Stability in the Mechanics of Continua | Schroeder, F.H. | Springer-Verlag | Berlin, Germany | 1982 | | | |
| 81-7 | 15 | High Temperature Gas Dynamics | Pichal, M. | Czech. Academy of Sciences | Prague, CSSR | | | | |
| 82-1 | 29 | Three-Dimensional Turbulent Boundary Layers | Fernholz, H.H. and Krause, E. | Springer-Verlag | Berlin, Germany | 1982 | | | |
| 82-2 | 51 | Giant Planets and Their Satellites | Kivelson, M.G. | | | 1983 | Advances in Space Research | 3 | 3 |
| 82-3 | | Impact Processes of Solid Bodies | McDonnell, J.A.M. | | | 1982 | Advances in Space Research | 2 | 12 |
| 82-4 | 29 | Fundamental Aspects of Material Sciences in Space | Malmejac, Y. | | | 1983 | Advances in Space Research | 3 | 5 |
| 82-5 | 58 | Modern Developments in Analytical Mechanics | Benenti, S., Francaviglia, M. and Lichnerowicz, A. | Academy of Sciences of Torino | Torino, Italy | 1983 | | | |
| 82-6 | 31 | Mechanics of Composite Materials | Hashin, Z., Herakovich, C.T. | Pergamon Press | New York, NY, USA | 1983 | | | |
| 82-7 | 69 | Nonlinear Deformation Waves | Nigul, U., Engelbrecht, J. | Springer-Verlag | Berlin, Germany | 1983 | | | |
| 82-8 | 30 | Collapse – The Buckling of Structures in Theory and Practice | Thompson, J.M.T. and Hunt, G.W. | Cambridge University Press | Cambridge, UK | 1983 | | | |
| 82-9 | 65 | Deformation and Failure of Granular Materials | Vermeer, P.A. and Luger, H.J. | A.A. Balkema | Rotterdam, Netherlands | 1982 | | | |
| 82-10 | 37 | Structure of Complex Turbulent Shear Flow | Dumas, R. and Fulachier, L. | Springer-Verlag | Berlin, Germany | 1983 | | | |
| 82-11 | 33 | Metallurgical Applications of Magnetohydro- dynamics | Moffatt, H.K. and Proctor, M.R.E. | | | 1983 | | | |
| 82-12 | 33 | Random Vibrations and Reliability | Hennig, K. | Academie-Verlag | Berlin, Germany | | | | |
| 83-1 | | Mechanical Behaviour of Electromagnetic Solid Continua | Maugin, G.A. | Elsevier Science Publishing Co. | New York, NY, USA | 1984 | | | |

| Symposia | Total No. Papers | Symposia Name | Editors of Books (or Special Issue of Journals) | Publisher | City | Year | Name | Volume No. | Issue No. |
|---|---|---|---|---|---|---|---|---|---|
| 83-2 | 34 | Measuring Techniques in Gas-Liquid Two-Phase Flows | Delhaye, J.M. and Cognet. G. | Springer-Verlag | Berlin, Germany | 1984 | | | |
| 83-3 | | Mechanics of Hearing | de Boer, E., Yiergever, M.A. | Martinus Nijhoff and Kluwer Academic | The Hague and New York | 1983 | | | |
| 83-4 | | Atmospheric Dispersion of Heavy Gases and Small Particles | Ooms, G. and Tennekcs, H. | Springer-Verlag | Berlin, Germany | 1984 | | | |
| 83-5 | | Seabed Mechanics | Denness, B. | Graham and Trotman | London, UK | 1984 | | | |
| 83-6 | | Turbulence and Chaotic Phenomena in Fluids | Tatsumi, T. | North- Holland Publishing Co. | Amsterdam, Netherlands | 1984 | | | |
| 83-7 | | Mechanics of Geomaterials: Rocks, Concretes, and Soils | Bazant, Z.P. and Rice, J.R. | John Wiley | Chichester, W. Sussex, UK and New York, USA | 1985 | | | |
| 84-1 | 47 | Fundamentals of Deformation and Fracture | | Cambridge Univ. Press | Cambridge, UK | 1984 | | | |
| 84-2 | 56 | Probabilistic Methods in the Mechanics of Solids and Structures | Eggwertz, S. and Lind, N.C. | Springer-Verlag | Berlin and New York | 1985 | | | |
| 84-3 | 37 | Influence of Polymer Additives on Velocity and Temperature Fields | Gampert, B. | Springer-Verlag | Berlin and New York | 1986 | | | |
| 84-4 | 96 | Laminar-Turbulent Transition | Kozlov, V.V. | Springer-Verlag | Berlin and New York | 1985 | | | |
| 84-5 | 48 | Optical Methods in Dynamics of Fluids and Solids | Pichal, M. | Springer-Verlag | Berlin and New York | 1985 | | | |
| 85-1 | 30 | Mechanics of Damage and Fatigue | | | | | Journal of Engineering Fracture Mechanics | | |

| Symposia | Total No. Papers | Symposia Name | Editors of Books (or Special Issue of Journals) | Publisher | City | Year | Name | Volume No. | Issue No. |
|---|---|---|---|---|---|---|---|---|---|
| 85-2 | 47 | Aero- and Hydro-Acoustics | Comte-Bellot, G. and Flowes - Williams, J.E. | Springer-Verlag | Berlin and New York | 1986 | | | |
| 85-3 | 35 | Hydrodynamics of Ocean Wave-Energy Utilization | Evans, D.V. and Falcao, A.F. de O. | Springer-Verlag | Berlin and New York | 1986 | | | |
| 85-4 | 31 | Inelastic Behaviour of Plates and Shells | Bevilacqua, L., Feijoo, R. and Valid, R. | Springer-Verlag | Berlin and New York | | | | | |
| 85-5 | 38 | Macro- and Micro-Mechanics of High Velocity Deformation and Fracture | | Springer-Verlag | Berlin and New York | | | | | |
| 85-6 | 50 | Mixing in Stratified Fluids | | | | 1987 | Journal of Geophysical Research | 92 | C5 |
| 85-7 | 32 | Turbulent Shear-Layer/Shock-Wave Interactions | Delery, J. | Springer-Verlag | Berlin and New York | 1986 | | | |
| 85-8 | 26 | Dynamics of Multi-body Systems | Bianchi, G., Schiehlen, W. | Springer-Verlag | Berlin and New York | 1986 | | | |
| 85-9 | | Simple and Multi-Phase Fluid Flow through Heterogeneollio Permeable Matenals | Donaldson, I.G. and Wooding, R.A. | | | | 1986 | Transport in Porous Media | 1 | |
| 86-1 | | Fluid Mechanics in the Spirit of G.I. Taylor | Batchelor, G.K | | | | 1986 | Journal of Fluid Dynamics | 173 | |
| 86-2 | | Boundary Layer Separation | Brown, S., Smith, S.T. | Springer-Verlag | Berlin | 1987 | | | |
| 86-3 | | Thermomechantcal Coupling in Solids | Bui, H.D., Nguyen, Q.S. | North-Holland Publishing Co. | Amsterdam | 1987 | | | |
| 86-4 | | Interaction in Deformable Solids and Structures | Yamamoto, Y. and Miya, K. | North-Holland Publishing Co. | Amsterdam | 1987 | | | |
| 86-5 | | Application of Asymptotic Methods to Problems in Continuum Mechanics | Cairlet, P.G. and Sanchez-Palencia, E. | Masson | Paris | 1987 | | | |

| Symposia | Total No. Papers | Symposia Name | Editors of Books (or Special Issue of Journals) | Publisher | City | Year | Name | Volume No. | Issue No. |
|---|---|---|---|---|---|---|---|---|---|
| 87-1 | | Turbulence Management and Relaminarisation | Marasimha, R. and Liepmann, H. | Springer-Verlag | Berlin | 1987 | ISBN 3-540-18574-7 | | |
| 87-2 | | Advanced Boundary Element Methods: Applications in Solid and Fluid Mechanics | Cruse, T.A. | Springer-Verlag | Berlin | 1987 | ISBN 3-540-17454-0 | | |
| 87-3 | | Nonlinear Stochastic Dynamic Engineering Systems | Ziegler, F., Schuëller, G. | Springer-Verlag | Berlin | 1988 | ISBN 3-540-18804-5 | | |
| 87-4 | | Yielding, Damage and Failure of Anisotropic Solids, Antoni Sawczuk in Memoriam | Boehler, J.P. | Mechanical Engineering Publ. Ltd | Bury St. Edmunds | 1990 | ISBN 0-852-98735-8 | | |
| 87-5 | | Nonlinear Acoustics | Kedrinskii, V.K. | Siberian Division | Novosibirsk | 1987 | | | |
| 87-6 | | Non-Linear Water Waves | Horikawa, K. and Maruo, H. | Springer-Verlag | Berlin | 1988 | ISBN 3-540-18793-6 | | |
| 87-7 | | Fundamental Aspects of Vortex Motion | Hasimoto, H. and Kambe, T. | North-Holland Publishing Co. | Amsterdam | 1988 | ISBN 0-444-87142-X | | |
| 88-1 | | Structural Optimization | Rozvany, G.I.N. and Karihaloo, B.L. | Kluwer Academic Publishers | Dordrecht | 1988 | ISBN 90-247-3771-0 | | |
| 88-2 | | Recent Advances in Nonlinear Fracture Mechanics | Knauss, W.G. and Rosakis, A.J. | | | 1990 | International Journal of Fracture | 42 | 1-4 |
| 88-3 | | Elastic Waves | McCarthy, M.F. and Hayes, M.A. | North-Holland, Elsevier Science Publ. | Amsterdam | 1989 | ISBN 0-444-87272-8 | | |
| 88-4 | | Liquid Metal Magnetohydrodynamics | Lielpeteris, J. and Moreau, R. | Kluwer Academic Publishers | Dordrecht | 1989 | ISBN 0-7923-0344-X | | |
| 88-5 | | Transsonicum III | Zierep, J. and Oertel, H. | Springer-Verlag | Berlin | 1989 | ISBN 3-540-50202-5 | | |

| Symposia | Total No. Papers | Symposia Name | Editors of Books (or Special Issue of Journals) | Publisher | City | Year | Name | Volume No. | Issue No. |
|---|---|---|---|---|---|---|---|---|---|
| 88-6 | | Dynamics of Controlled Mechanical Systems | Schweitzer, G. and Mansour, M. | Springer-Verlag | Berlin | 1989 | ISBN 3-540-50201-7 | | |
| 89-1 | | Discretization Methods in Structural Mechanics | Kuhn, G. and Mang, H. | Springer-Verlag | Berlin | 1990 | ISBN 3-540-52011-2 | | |
| 89-2 | | Structure of Turbulence and Drag Reduction | Gyr, A. | Springer-Verlag | Berlin | 1990 | ISBN 3-540-50204-1 | | |
| 89-3 | | Elastic Wave Propagation and Ultrasonic Nondestructive Evaluation | Datta, S.K., Achenbach, J.D. and Rajapakse, Y.S. | North-Holland | Amsterdam | 1990 | ISBN 0-444-87485-2 | | |
| 89-4 | | Topological Fluid Mechanics | Moffat, H.K. and Tsinober, A. | Cambridge University Press | Cambridge | 1990 | ISBN 0-521-38145-2 | | |
| 89-5 | | Ice-Structure Interaction | Jones, S.J., McKenna, R.S., Tillotson, J., Jordaan, I.J. | Springer-Verlag | Berlin | 1991 | ISBN 3-540-52192-5 | | |
| 89-6 | | Nonlinear Dynamics in Engineering Systems | Schiehlen, W. | Springer-Verlag | Berlin | 1990 | ISBN 3-540-50200-9 | | |
| 89-7 | | Adiabatic Waves in Liquid-Vapor Systems | Meier, G.E.A. and Thompson, P.A. | Springer-Verlag | Berlin | 1990 | ISBN 3-540-50203-3 | | |
| 89-8 | | Laminar-Turbulent Transition | Arnal, D., Michel, R. | Springer-Verlag | Berlin | 1990 | ISBN 3-540-52196-8 | | |
| 90-1 | | Mechanical Modellings of New Electromagnetic Materials | Hsieh, R.K.T | Elsevier | Amsterdam | 1990 | ISBN 0-444-88518-8 | | |
| 90-2 | | Dynamical Problems of Rigid-Elastic Systems and Structures | Banichuk, N.V., Klimov, D.M. and Schiehlen, W. | Springer-Verlag | Berlin | 1991 | ISBN 3-540-53788-0 | | |

| Symposia | Total No. Papers | Symposia Name | Editors of Books (or Special Issue of Journals) | Publisher | City | Year | Name | Volume No. | Issue No. |
|---|---|---|---|---|---|---|---|---|---|
| 90-3 | | Inelastic Deformation of Composite Materials | Dvorak, G.J. | Springer-Verlag | New York | 1991 | ISBN 0-387-97458-X | | |
| 90-4 | | Dynamics of Marine Vehicles and Structures in Waves | Price, W.G., Temarel, P., Keane, A.J. | Elsevier | Amsterdam | 1991 | ISBN 0-444-89020-3 | | |
| 90 5 | | Separated Flows and Jets | Kozlov, V.V., Dovgal, A.V. | Springer-Verlag | Berlin | 1991 | ISBN 3-540-53762-7 | | |
| 90-6 | | Fluid Mechanics of Stirring and Mixing | Acrivos, A. | | | 1991 | Physics of Fluids A | 3 | |
| 90-7 | | Nonlinear Hydrodynamics Stability and Transition | Iooss, G. | | | 1991 | European Journal of Mechanics B/Fluids | 10 | 2 |
| 90-8 | | Contact Load and Local Effects in Thin-Walled Plated and Shell Structures | Křupka, V. and Drdácký, M. | Springer-Verlag | Berlin | 1992 | ISBN 3-540-53551-9 | | |
| 90-9 | | Creep in Structures | Życzkowski, M. | Springer-Verlag | Berlin | 1992 | ISBN 3-540-53786-4 | | |
| 91-1 | | Aerothermodynamics in Combusters | Lee, R.S.L., Whitelaw, J.H. and Wung, T.S. | Springer-Verlag | Berlin | 1992 | ISBN 3-540-55404-1 | | |
| 91-2 | | Mechanical Effects of Welding | Karlsson, L., Lindgren, L.E. and Jonsson, M. | Springer-Verlag | Berlin | 1992 | ISBN 3-540-55240-5 | | |
| 91-3 | | Nonlinear Stochastic Mechanics | Bellomo, N. and Casciati, F. | Springer-Verlag | Berlin | 1992 | ISBN 3-540-55545-5 | | |
| 91-4 | | Mechanics of Fluidized Beds | Homsy, G.M., Jackson, R. and Grace, J.R. | | | 1992 | Journal of Fluid Mechanics | 236 | |
| 91-5 | | Breaking Waves | Banner, M.L., and Grimshaw, R. | Springer-Verlag | Berlin | 1992 | ISBN 3-540-55944-2 | | |

| Symposia | Total No. Papers | Symposia Name | Editors of Books (or Special Issue of Journals) | Publisher | City | Year | Name | Volume No. | Issue No. |
|---|---|---|---|---|---|---|---|---|---|
| 91-6 | | Constitutive Relations for Finite Deformation of Polycrystalline Metals | Ren Wang and Drucker, D.C. | Springer-Verlag | Berlin | 1992 | ISBN 3-540-55128-X | | |
| 91-7 | | Finite Inelastic Deformations - Theory and Application | Besdo, D. and Stein, E. | Springer-Verlag | Berlin | 1992 | ISBN 3-540-55849-7 | | |
| 91-8 | | Interpretation of Time Series from Mechanical Systems | Drazin, P.G. and King, G.P. | North Holland, Elsevier Science Publ | Amsterdam | 1992 | Physica D Nonlinear Phenomena | 58 | |
| 91-9 | | Microgravity Fluid Mechanics | Rath, H.J. | Springer-Verlag | Berlin | 1992 | ISBN 3-540-55122-0 | | |
| 91-10 | | Local Mechanics Concepts for Composite Material Systems | Reddy, J.N. and Reifsnider, K.L. | Springer-Verlag | Berlin | 1992 | ISBN 3-540-55547-1 | | |
| 92-1 | | Optimal Control of Mechanical Systems | Chernousko, F.L. | English translation published by Scripta Technica Inc., A. Wiley Company | New York | 1993 | Izvestiya of the Russian Academy of Sciences, Tekhnicheskaya Kibernetika | | 1 |
| 92-2 | | Inverse Problems in Engineering Mechanics | Tanaka, M. and Bui, H.D. | Springer-Verlag | Berlin | 1993 | ISBN 3-540-56345-8 | | |
| 92-3 | | Optimal Design with Advanced Materials (Niordson Anniversary) | Pedersen, P. | Elsevier Science Publishers | Amsterdam | 1993 | ISBN 0444-89869-7 | | |
| 92-4 | | Aerothermochemistry of Spacecraft and Associated Hypersonic Flows | Brun, R. and Chikhaoui, A.A. | Jouve, 18, rue Saint-Denis | Paris | 1994 | No. 215515N | | |
| 92-5 | | Bluff-Body Wakes, Dynamics and Instabilities | Eckelmann, H., Graham, J.M.R., Huerre, P., Monkewitz, P.A. | Springer-Verlag | Berlin | 1993 | ISBN 3-540-56594-9 | | |

| Symposia | Total No. Papers | Symposia Name | Editors of Books (or Special Issue of Journals) | Publisher | City | Year | Name | Volume No. | Issue No. |
|---|---|---|---|---|---|---|---|---|---|
| 92-6 | | Fluid Dynamics of High Angle of Attack | Kawamura, R. and Aihara, Y. | Springer-Verlag | Berlin | 1993 | ISBN 3-540-56593-0 | | |
| 92-7 | | Eddy Structure Identification in Free Turbulent Shear Flow | Bonnet, J.P. and Glauser, M.N. | Kluwer Academic Publishers | Dordrecht | 1993 | ISBN 0-7923-2449-8 | | |
| 93-1 | | Probabilistic Structural Mechanics: Advances in Structural Reliability Methods | Spanos, P.D. and Wu, Y.-T. | Springer-Verlag | Berlin | 1994 | ISBN 3-540-57709-2 | | |
| 93-2 | | Computational Mechanics of Materials | Ortiz, M. and Shih, C.F. | | | 1994 | Modelling and Simulation in Materials Science and Engineering | 2 | 3A |
| 93-3 | | Nonlinearity and Chaos in Engineering Dynamics | Thompson, J.M.T. and Bishop, S.R. | John Wiley & Sons | Chichester, UK | 1994 | ISBN 0-471-94458-0 | | |
| 93-4 | | Nonlinear Instability of Nonparallel Flows | Lin, S.P., Phillips, W.R.C. and Valentine, D.T. | Springer-Verlag | Berlin | 1994 | ISBN 3-540-57679-7. | | |
| 93-5 | | Nonlinear Waves and Solids | Wegner, J.L. and Norwood, F.R. | The American Society of Mechanical Engineers | New York | 1995 | ISBN 0-7918-0645-6 | | |
| 93-6 | | Identification of Mechanical Systems | Müller, P.C. | Springer-Verlag | Berlin | 1995 | | | |
| 93-7 | | Discrete Structural Optimization | Gutkowski, W. | Springer-Verlag | Berlin | 1994 | ISBN 3-540-57679-X | | |
| 93-8 | | Bubble Dynamics and Interface Phenomena | Blake, J.R., Boulton-Stone, J.M. and Thomas, N.H. | Kluwer Academic Publishers | Dordrecht | 1994 | ISBN 0-7923-3008-0 | | |
| 93-9 | | Fracture of Brittle Disordered Materials: Concrete, Rock and Ceramics | Baker, G. and Karihaloo, B.L. | E & FN Spon | London | 1995 | ISBN 0-419-19050-3 | | |

| Symposia | Total No. Papers | Symposia Name | Editors of Books (or Special Issue of Journals) | Publisher | City | Year | Name | Volume No. | Issue No. |
|---|---|---|---|---|---|---|---|---|---|
| 93-10 | | Impact Dynamics | Cheng, C.-M. and Tan, Q. | Peking University Press | Beijing | 1994 | ISBN 7-301-02489-4/0 338 | | |
| 93-11 | | Numerical Simulation of Nonisothermal Flow of Viscoelastic Liquids | Dijksman, J.F. and Kuiken, G.D.C. | Kluwer Academic Publishers | Dordrecht | 1995 | ISBN 0-7923-3262-8 | | |
| 94-1 | | no proceedings | | | | | | | |
| 94-2 | | Waves and Liquid/Gas and Liquid/Vapor Two-Phase Systems | Morioka, S. and van Wijngaarden, L. | Kluwer Academic Publishers | Dordrecht | 1995 | ISBN 0-7923-3424-8 | | |
| 94-3 | | Structure and Dynamics of Nonlinear Waves in Fluids | Mielke, A. and Kirchgässner, K. | World Scientific Publishing | Singapore | 1995 | ISBN 981-02-2124-X | | |
| 94-4 | | Microstructure-Property Interactions in Composite Materials | Pyrz, R. | Kluwer Academic Publishers | Dordrecht | 1995 | ISBN 0-7923-3427-2 | | |
| 94-5 | | Anisotopy, Inhomogeneity and Nonlinearity in Solid Mechanics | Parker, D.F. and England, A.H. | Kluwer Academic Publishers | Dordrecht | 1995 | ISBN 0-7923-3594-5 | | |
| 94-6 | | Laminar-Turbulent Transition | Kobayashi, R. | Springer-Verlag | Berlin | 1995 | ISBN 3-540-59297-0 | | |
| 94-7 | | Mechanical Problems in Geodynamics | Wang, R. and Aki, K. | | | 1996 | Pure and Applied Geophysics, ISSN 0033-4553, ISBN 3-7643-5412-7 | 145-146 | 3/4 |
| 94-8 | | The Active Control of Vibration | Burrows, C.R. and Keogh, P.S. | Mechanical Engineering Publications Limited | London | 1994 | ISBN 0-85298-916-4 | | |
| 94-9 | | Size-Scale Effects in the Failure Mechanisms of Materials and Structures | Carpinteri, A. | E & FN Spon | London | 1995 | ISBN 0-419-20520-9 | | |

| Symposia | Total No. Papers | Symposia Name | Editors of Books (or Special Issue of Journals) | Publisher | City | Year | Name | Volume No. | Issue No. |
|---|---|---|---|---|---|---|---|---|---|
| 94-10 | | Mechanics and Combustion of Droplets and Sprays | Chiu, H.H. | Begell House Publishers | New York | 1995 | ISBN 1-56700-051-7 | | |
| 95-1 | | Optimization of Mechanical Systems | Bestle, D., Schiehlen, W. | Kluwer Academic Publishers | Dordrecht | 1996 | ISBN 0-7923-3830-8 | | |
| 95-2 | | Asymptotic Methods for Turbulent Shear Flows at High Reynolds Numbers | Gersten, K. | Kluwer Academic Publishers | Dordrecht | 1996 | ISBN 0-7923-4138-4 | | |
| 95-3 | | Advances in Nonlinear Stochastic Mechanics | Naess, A. and Krenk, S. | Kluwer Academic Publishers | Dordrecht | 1996 | ISBN 0-7923-4193-7 | | |
| 95-4 | | Nonlinear Instability and Transition in 3D Boundary Layers | Duck, P.W. and Hall, P. | Kluwer Academic Publishers | Dordrecht | 1996 | ISBN 0-7923-4079-5 | | |
| 95-5 | | no proceedings | | | | | | | |
| 95-6 | | Micromechanics of Plasticity and Damage of Multiphase Materials | Pineau, A. and Zaoui, A. | Kluwer Academic Publishers | Dordrecht | 1996 | ISBN 0-7923-4188-0 | | |
| 95-7 | | Nonlinear Analysis of Fracture | Willis, J. | Kluwer Academic Publishers | Dordrecht | 1996 | ISBN 0-7923-4378-6 | | |
| 95-8 | | no proceedings | | | | | | | |
| 95-9 | | Combustion in Supersonic Flows | Champion, M. and Deshales, B. | Kluwer Academic Publishers | Dordrecht | 1997 | ISBN 0-7923-4313-1 | | |
| 95-10 | | no proceedings | | | | | | | |
| 96-1 | | Interaction between Dynamics and Control in Advanced Mechanical Systems | van Campen, D.H. | Kluwer Academic Publishers | Dordrecht | 1997 | ISBN 0-7923-4429-4 | | |
| 96-2 | | Innovative Computational Methods for Fracture and Damage | O'Donoghue, P., Gilchrist, M., Broberg, K.B. | | | 1997 | Computational Mechanics Journal | 19-20 | |
| 96-3 | | Variable Density Low Speed Turbulent Flows | Fulachier, L., Lumley, J.L. and Anselmet, F. | Kluwer Academic Publishers | Dordrecht | 1997 | ISBN 0-7923-4602-5 | | |

| Symposia | Total No. Papers | Symposia Name | Editors of Books (or Special Issue of Journals) | Publisher | City | Year | Name | Volume No. | Issue No. |
|---|---|---|---|---|---|---|---|---|---|
| 96-4 | | Mechanics of Granular and Porous Materials | Fleck, N.A. and Cocks, A.C.F. | Kluwer Academic Publishers | Dordrecht | 1997 | ISBN 0-7923-4553-3 | | |
| 97-1 | | Lubricated Transport of Viscious Materials | Ramkissoon, H. | Kluwer Academic Publishers | Dordrecht | 1997 | ISBN 0-7923-4897-4 | | |
| 97-2 | | Transformation Problems in Composite and Active Materials | Bahei-El-Din, Y.A. and Dvorak, G.J. | Kluwer Academic Publishers | Dordrecht | 1998 | ISBN 0-7923-5122-3 | | |
| 97-3 | | Non-Linear Singularities in Deformation and Flow | Durban, D. and Pearson, J.R.A. | Kluwer Academic Publishers | Dordrecht | 1998 | ISBN 0-7923-5349-8 | | |
| 97-4 | | Variations of Domains and Free-boundary Problems in Solid Mechanics | Argoul, P., Frémond, M. and Nguyen, Q.S. | Kluwer Academic Publishers | Dordrecht | 1998 | ISBN 0-7923-5450-8 | | |
| 97-5 | | Simulation and Identification of Organized Structures in Flows | Sørensen, J.N., Hopfinger, E.J. and Aubry, N. | Kluwer Academic Publishers | Dordrecht | 1999 | ISBN 0-7923-5603-9 | | |
| 97-6 | | Discretization Methods in Structural Mechanics II | Mang, H.A. and Rammerstorfer, F.G. | Kluwer Academic Publishers | Dordrecht | 1999 | ISBN 0-7923-5591-1 | | |
| 97-7 | | Material Instabilities in Solids | de Borst, R. and van der Giessen, E. | John Wiley & Sons | Chichester, UK | 1998 | ISBN 0-471-97460-9 | | |
| 97-8 | | Statistical Energy Analysis | Fahy, F.J. and Price, W.G | Kluwer Academic Publishers | Dordrecht | 1998 | ISBN 0-7923-5457-5 | | |
| 97-9 | | Rheology and Computation | | | | 1999 | Journal of Non-Newtonian Fluid Mechanics | several volumes | |
| 97-10 | | New Applications of Nonlinear and Chaotic Dynamics in Mechanics | Moon, F.C. | Kluwer Academic Publishers | Dordrecht | 1998 | ISBN 0-7923-5276-9 | | |

| Symposia | Total No. Papers | Symposia Name | Editors of Books (or Special Issue of Journals) | Publisher | City | Year | Name | Volume No. | Issue No. |
|---|---|---|---|---|---|---|---|---|---|
| 97-11 | | Computational Methods for Unbounded Domains | Geers, T.L. | Kluwer Academic Publishers | Dordrecht | 1998 | ISBN 0-7923-5266-1 | | |
| 97-12 | | Micro- and Macrostructural Aspects of Thermoplasticity | Bruhns, O.T. and Stein, E. | Kluwer Academic Publishers | Dordrecht | 1998 | ISBN 0-7923-5265-3 | | |
| 97-13 | | Dynamics of Slender Vortices | Krause, E. and Gersten, K. | Kluwer Academic Publishers | Dordrecht | 1998 | ISBN 0-7923-5041-3 | | |
| 97-14 | | Rheology of Buddies with Defects | Wang, R. | Kluwer Academic Publishers | Dordrecht | 1998 | ISBN 0-7923-5297-1 | | |
| 98-1 | | Three-Dimensional Aspects of Air-Sea Interaction | Dias, F. and Khariff, C. | | | 1999 | European Journal of Mechanics B / Fluids | 18 | 3 |
| 98-2 | | Synthesis in Bio Solid Mechanics | Pedersen, P., Bendsøe, M.P. | Kluwer Academic Publishers | Dordrecht | 1999 | ISBN 0-7923-5615-2 | | |
| 98-3 | | Developments in Geophysical Turbulence | Kerr, R.M. and Kimura, Y. | Kluwer Academic Publishers | Dordrecht | 2000 | ISBN 0-7923-6673-5 | | |
| 98-4 | | Viscoelastic Fluid Mechanics: Effects of Molecular Modelling | Shaqfeh, E.S.G. | | | 1999 | Journal of Non-Newtonian Fluid Mechanics | 82 | |
| 98-5 | | Unilateral Multibody Dynamics | F. Pfeiffer and Ch. Glocker | Kluwer Academic Publishers | Dordrecht | 1999 | ISBN 0-7923-6030-3 | | |
| 98-6 | | Synthesis of Nonlinear Dynamical Systems | Lavendelis, E. and Zakrzhevsky, M. | Kluwer Academic Publishers | Dordrecht | 1999 | ISBN 0-7923-6106-7 | | |
| 98-7 | | Advanced Optical Methods and Applications in Solid Mechanics | Lagarde, A. | Kluwer Academic Publishers | Dordrecht | 2000 | ISBN 0-7923-6604-2 | | |
| 98-8 | | Deployable Structures: Theory and Applications | Pellegrino, S. and Guest, S.D. | Kluwer Academic Publishers | Dordrecht | 2000 | ISBN 0-7923-6516-X | | |

| Symposia | Total No. Papers | Symposia Name | Editors of Books (or Special Issue of Journals) | Publisher | City | Year | Name | Volume No. | Issue No. |
|---|---|---|---|---|---|---|---|---|---|
| 98-9 | | Mechanics of Passive & Active Flow Control (FLOWCON) | Meier, G.E.A. and Viswanath, P.R. | Kluwer Academic Publishers | Dordrecht | 1999 | ISBN 0-7923-5928-3 | | |
| 99-1 | | Nonlinearity and Stochastic Structural Dynamics | Narayanan, S., Iyengar, R.N. | Kluwer Academic Publishers | Dordrecht | 2000 | ISBN 0-7923-6733-2 | | |
| 99-2 | | Mechanical and Electromagnetic Waves in Structured Media | McPhedran, R.C., Botten, L.C., Nicorovici, N.A. | Kluwer Academic Publishers | Dordrecht | 2001 | ISBN 0-7923-7038-4 | | |
| 99-3 | | Recent Developments in Non-linear Oscillations of Mechanical Systems | Van Dao, N. and Kreuzer, E.J. | Kluwer Academic Publishers | Dordrecht | 2000 | ISBN 0-7923-6470-8 | | |
| 99-4 | | Advanced Mathematical and Computational Mechanics Aspects of the Boundary Element Method | Burczynski, T. | Kluwer Academic Publishers | Dordrecht | 2001 | ISBN 0-7923-7081-3 | | |
| 99-5 | | Segregation in Granular Flows | Rosato, A.D. and Blackmore, D.L. | Kluwer Academic Publishers | Dordrecht | 2000 | ISBN 0-7923-6547-X | | |
| 99-6 | | Nonlinear Waves in Multi-Phase Flow | Chang, H.C. | Kluwer Academic Publishers | Dordrecht | 2000 | ISBN 0-7923-6454-6 | | |
| 99-7 | | Theoretical and Numerical Methods in Continuum Mechanics of Porous Materials | Ehlers, W. | Kluwer Academic Publishers | Dordrecht | 2001 | ISBN 0-7923-6766-9 | | |
| 99-8 | | Laminar-Turbulent Transition | Fasel, H. and Saric, W.S. | Springer-Verlag | Berlin, Heideberg, New York | 2000 | ISBN 3-540-67947-2 | | |
| 99-9 | | Geometry and Statistics of Turbulence | Kambe, T., Nakano, T. and Miyauchi, T. | Kluwer Academic Publishers | Dordrecht | 2001 | ISBN 0-7923-6711-1 | | |
| 00-1 | | Creep in Structures | Murakami, S. and Ohno, N. | Kluwer Academic Publishers | Dordrecht | 2000 | ISBN 0-7923-6737-5 | | |

| Symposia | Total No. Papers | Symposia Name | Editors of Books (or Special Issue of Journals) | Publisher | City | Year | Name | Volume No. | Issue No. |
|---|---|---|---|---|---|---|---|---|---|
| 00-2 | | Bluff Body Wakes and Vortex-Induced Vibrations | Leweke, T., Bearman, P.W. and Williamson, C.H.K. | | | 2001 | Journal of Fluids and Structures | 15 | 3/4 |
| 00-3 | | Mechanical Waves for Composite Structures Characterization | Dempsey, J.P. and Shen, H.H. | Kluwer Academic Publishers | Dordrecht | 2001 | ISBN 1-4020-0171-1 | | |
| 00-4 | | Advances in Mathematical Modelling of Atmosphere and Ocean Dynamics | Sotiropoulos, D.A | Kluwer Academic Publishers | Dordrecht | 2001 | ISBN 0-7923-7164-X | | |
| 00-5 | | Free Surface Flows | Hodnett, P.F. | Kluwer Academic Publishers | Dordrecht | 2001 | ISBN 0-7923-7075-9 | | |
| 00-6 | | Diffraction and Scattering in Fluid Mechanics and Elasticity | King, A.C. and Shikhmurzaev, Y.D. | Kluwer Academic Publishers | Dordrecht | 2001 | ISBN 0-7923-7085-6 | | |
| 00-7 | | Field Analysis for Determination of Material Parameters Experimental and Numerical Aspects | Abrahams, I.D., Martin, P.A. and Simon, M.J. | Kluwer Academic Publishers | Dordrecht | 2002 | ISBN 1-4020-0590-3 | | |
| 00-8 | | Smart Structures and Structronic Systems | Gabbert, U. and Tzou, H.S. | Kluwer Academic Publishers | Dordrecht | 2001 | ISBN 0-7923-6968-8 | | |
| 00-9 | | Designing for Quietness | Munjal, M.L. | Kluwer Academic Publishers | Dordrecht | 2002 | ISBN 1-4020-0765-5 | | |
| 01-1 | | Flow in Collapsible Tubes and Past Other Highly Compliant Boundaries | Carpenter, P.W. and Pedley, T.J. | Kluwer Academic Publishers | Dordrecht | 2003 | ISBN 1-4020-1161-X | | |
| 01-2 | | Material Instabilities and the Effect of Microstructure | Kyriakides, S., Triantafyllidis, N. | | | 2002 | International Journal of Solids and Structures | 39 | 13-14 |
| 01-3 | | Turbulent Mixing and Combustion | Pollard, A. and Candel, S. | Kluwer Academic Publishers | Dordrecht | 2002 | ISBN 1-4020-0747-7 | | |

| Symposia | Total No. Papers | Symposia Name | Editors of Books (or Special Issue of Journals) | Publisher | City | Year | Name | Volume No. | Issue No. |
|---|---|---|---|---|---|---|---|---|---|
| 01-4 | | Micromechanics of Martensitic Phase Transformation in Solids | Sun, Q.P. | Kluwer Academic Publishers | Dordrecht | 2002 | ISBN 1-4020-0741-8 | | |
| 01-5 | | Analytical and Computational Fracture Mechanics of Non-homogenous Materials | Karihaloo, B.L. | Kluwer Academic Publishers | Dordrecht | 2002 | ISBN 1-4020-0510-5 | | |
| 01-6 | | Computational Mechanics of Solid Materials at Large Strains | Miehe, C. | Kluwer Academic Publishers | Dordrecht | 2003 | ISBN 1-4020-1170-9 | | |
| 01-7 | | Tubes, Sheets and Singularities in Fluid Dynamics | Bajer, K. and Moffatt, H.K. | Kluwer Academic Publishers | Dordrecht | 2002 | ISBN 1-4020-0980-1 | | |
| 02-1 | | Micromechanics of Fluid Suspensions and Solid Composites | | | | 2003 | Philosophical Transactions: Mathematical, Physical & Engineering Sciences | | |
| 02-2 | | Unsteady Separated Flows | Braza, M. Hirsch, C., Hussain, F. | | | 2003 | Flow, Turbulence and Combustion | 71 | 1-4 |
| 02-3 | | Dynamics of Advanced Materials and Smart Structures | Watanabe, K. and Ziegler, F. | Kluwer Academic Publishers | Dordrecht | 2003 | ISBN 1-4020-1061-3 | | |
| 02-4 | | Asymptotics, Singularities and Homogenisation in Problems of Mechanics | Movchan, A.B. | Kluwer Academic Publishers | Dordrecht | 2003 | ISBN 1-4020-1780-4 | | |
| 02-5 | | Complementary-Dual-Varational Principles in Nonlinear Mechanics | Gao, D.Y. | Kluwer Academic Publishers | Dordrecht | 2004 | ISBN 1-4020-7887-0 (HB) and ISBN 1-4020-7888-9 (E-book) | | |
| 02-6 | | Nonlinear Stochastic Systems | Namachchivaya, S., Lin, Y.K. | Kluwer Academic Publishers | Dordrecht | 2003 | ISBN 1-4020-1471-6 | | |

| Symposia | Total No. Papers | Symposia Name | Editors of Books (or Special Issue of Journals) | Publisher | City | Year | Name | Volume No. | Issue No. |
|---|---|---|---|---|---|---|---|---|---|
| 02-7 | | Transsonicum IV | Sobieczky, H. | Kluwer Academic Publishers | Dordrecht | 2003 | ISBN 1-4020-1608-5 | | |
| 02-8 | | Reynolds Number Scaling in Turbulent Flow | Smits, A.J. | Kluwer Academic Publishers | Dordrecht | 2003 | ISBN 1-4020-1775-8 | | |
| 02-9 | | Evolutionary Methods in Mechanics | Burczynski, T. and Osyczka, A. | Kluwer Academic Publishers | Dordrecht | 2004 | ISBN 1-4020-2266-2 (HB) and ISBN 1-4020-2267-0 (E-book) | | |
| 02-10 | | Multiscale Modeling and Characterizzation of Elastic-Inelastic Behaviour of Engineering Materials | Ahzi, S., Charkaoui, M., Khaleel, M.A., Zbib, H.M., Zikry, M.A., LaMatina, B. | Kluwer Academic Publishers | Dordrecht | 2003 | ISBN 1-4020-1861-4 | | |
| 03-1 | | Mechanics of Physiochemical and Electromechanical Interactions in Porous Media | Huyghe, J.M., Raats, P.A.C. and Cowin, S.C. | Springer | Dordrecht | 2006 | ISBN 978-1-4020-3864-8 | | |
| 03-2 | | Integrated Modeling of Fully Coupled Fluid-Structure Interactions | Benaroya, H. and Wei, T. | Kluwer Academic Publishers | Dordrecht | 2003 | ISBN 1-4020-1806-1 | | |
| 03-3 | | Chaotic Dynamics and Control of Systems and Processes in Mechanics | Rega, G. and Vestroni, F. | Springer | Dordrecht | 2005 | ISBN 1-4020-3267-6 (HB) and ISBN 1-4020-3268-4 (E-book) | | |
| 03-4 | | Mesoscopic Dynamics of Fracture Process and Materials Strength | Kitagawa, H., Shibutani, Y. | Kluwer Academic Publishers | Dordrecht | 2004 | ISBN 1-4020-2037-6 (HB) and ISBN 1-4020-2111-9 116 Report 2013 (e-book) | | |

| Symposia | Total No. Papers | Symposia Name | Editors of Books (or Special Issue of Journals) | Publisher | City | Year | Name | Volume No. | Issue No. |
|---|---|---|---|---|---|---|---|---|---|
| 04-1 | | Size Effects on Material and Structural Behavior at Micron- and Nanometer- Scales | Sun, Q.P. and Tong, P. | Springer | Dordrecht | 2006 | ISBN 1-4020-4945-5 | | |
| 04-2 | | no proceedings | | | | | | | |
| 04-3 | | Size Effects on Non-Uniqueness of Solutions to the Navier-Stokes Equations and their Connection with Laminar-Turbulent Transition | Mullin, T. and Kerswell, R.R. | Springer | Dordrecht | 2005 | ISBN 1-4020-4048-2 | | |
| 04-4 | | One Hundred Years of Boundary Layer Research | Meier, G.E.A., Sreenivasan, K.R., et al. | Springer | Dordrecht | 2006 | ISBN 1-4020-4149-7 | | |
| 04-5 | | Elastohydrodynamics and Microelastohydrodynamics | Snidle, R.W., and Evans, H.P. | Springer | Dordrecht | 2006 | ISBN 1-4020-4532-8 | | |
| 04-6 | | Mechanics and Reliability of Actuating Materials | Yang, W. | Springer | Dordrecht | 2005 | ISBN 1-4020-4130-6 | | |
| 04-7 | | Recent Advances in Disperse Multiphase Flow Simulation | Balachandar, S., Prosperetti, A. | Springer | Dordrecht | 2006 | ISBN 1-4020-4976-5 | | |
| 04-8 | | Elementary Vortices and Coherent Structures: Significance in Turbulence Dynamics | Kida, S. | Springer | Dordrecht | 2006 | ISBN 1-4020-4180-2 | | |
| 04-9 | | Laminar-Turbulent Transition | Govindarajan, R. | Springer | Dordrecht | 2006 | ISBN 1-4020-3459-8 | | |
| 05-1 | | Multiscale Modelling of Damage and Fracture Processes in Composite Materials | Sadowski, T. | Springer Academic Publishers | Dordrecht | 2006 | ISBN 978-1-4020-4565-3 | | |

| Symposia | Total No. Papers | Symposia Name | Editors of Books (or Special Issue of Journals) | Publisher | City | Year | Name | Volume No. | Issue No. |
|---|---|---|---|---|---|---|---|---|---|
| 05-2 | | Mechanical Behavior and Micromechanics of Nanostructured Materials | Bai, Y.L., Zheng, Q.S. and Wei, Y.G. | Springer Academic Publishers | Dordrecht | 2007 | ISBN 978-1-4020-5623-9 | | |
| 05-3 | | Impact Biomechanics - From Fundamental Insights to Applications | Gilchrist, M.D. | Springer Academic Publishers | Dordrecht | 2005 | ISBN 978-1-4020-3795-5 | | |
| 05-4 | | Vibration Control of Nonlinear Mechanisms and Structures | Ulbrich, H. and Günthner, W. | Springer Academic Publishers | Dordrecht | 2005 | ISBN 978-1-4020-4160-0 | | |
| 05-5 | | Topological Design Optimization of Structures, Machines and Materials-Status and Perspectives | Bendsøe, M.P., Olhoff, N. and Sigmund, O. | Springer Academic Publishers | Dordrecht | 2006 | ISBN 978-1-4020-4729-9 | | |
| 06-1 | | Multiscale Problems in Multibody System Contacts | Eberhard, P. | Springer Academic Publishers | Dordrecht | 2006 | ISBN 978-1-4020-5980-3 | | |
| 06-2 | | Interactions for Dispersed Systems in Newtonian and Viscoelastic Fluids | | | | 2006 | Physics of Fluids | 18 | 121501-1 |
| 06-3 | | Plasticity at the Micron Scale | Tvergaard, V. | | | 2007 | Modelling and Simulation in Materials Science and Engineering | 15 | 1 |
| 06-4 | | Hamiltonian Dynamics, Vortex Structures, Turbulence | Borisov, A.V., Kozlov, V.V., et al. | Springer Academic Publishers | Dordrecht | 2007 | ISBN 978-1-4020-6743-3 | | |
| 06-5 | | Discretization Methods for Evolving Discontinuities | Combescure, A., de Borst, R., Belytschko, T. | Springer Academic Publishers | Dordrecht | 2007 | ISBN 978-1-4020-6529-3 | | |

| Symposia | Total No. Papers | Symposia Name | Editors of Books (or Special Issue of Journals) | Publisher | City | Year | Name | Volume No. | Issue No. |
|---|---|---|---|---|---|---|---|---|---|
| 06-6 | | Computational Physics and New Perspectives in Turbulence | Kaneda, Y. | Springer Academic Publishers | Dordrecht | 2007 | ISBN 978-1-4020-6471-5 | | |
| 06-7 | | Dynamics and Control of Nonlinear Systems with Uncertainty | Hu, H.Y., Kreuzer E.J. | Springer Academic Publishers | Dordrecht | 2007 | ISBN 978-1-4020-6331-2 | | |
| 06-8 | | Flow Control with Mems | Morrison, J.F., Birch, D.M., Lavoie, P. | Springer Academic Publishers | Dordrecht | 2007 | ISBN 978-1-4020-6857-7 | | |
| 06-9 | | Computational Contact Mechanics | Wriggers, P., Nackenhorst, U. | Springer Academic Publishers | Dordrecht | 2007 | ISBN 978-1-4020-6404-3 | | |
| 07-1 | | Relation of Shell, Plate, Beam and 3D Models | Jaiani, G. and Podio-Guidugli, P. | Springer Academic Publishers | Dordrecht | 2008 | ISBN 978-1-4020-8773-8 | | |
| 07-2 | | Recent Advances in Multiphase Flows: Numerical and Experimental | | | | 2008 | Physics of Fluids | 20 | 4 |
| 07-3 | | Unsteady Separated Flows and their Control | Braza, M. and Hourigan, K. | | | 2009 | Journal of Fluids and Structures | 24 | 8 |
| 07-4 | | Scaling in Solid Mechanics | Borodich, F. | Springer Academic Publishers | Dordrecht | 2008 | ISBN 978-1-4020-9032-5 | | |
| 07-5 | | Fluid-Structure Interaction in Ocean Engineering | Kreuzer, E. | Springer Academic Publishers | Dordrecht | 2009 | ISBN 978-1-4020-8629-8 | | |
| 07-6 | | Swelling and Shrinking of Porous Materials: From Colloid Science to Pro-Mechanics | | | | 2010 | Anais da Academia Brasileira de Ciencia | 82 | 1 |
| 07-7 | | Advances in Micro- and Nanofluidics | Ellero, M., Hu, X., Fröhlich, J., Adams, N. | Springer Academic Publishers | Dordrecht | 2009 | ISBN 978-90-481-2625-5 | | |

| Symposia | Total No. Papers | Symposia Name | Editors of Books (or Special Issue of Journals) | Publisher | City | Year | Name | Volume No. | Issue No. |
|---|---|---|---|---|---|---|---|---|---|
| 07-8 | | Mechanical Properties of Cellular Materials | Zhao, H., Fleck, N.A. | Springer Academic Publishers | Dordrecht | 2009 | ISBN 978-1-4020-9403-3 | | |
| 07-9 | | Multi-Scale Plasticity of Crystalline Materials | | | | 2008 | Philosophical Magazine | 88 | 30-32 |
| 08-1 | | Theoretical, Modelling and Computational Aspects of Inelastic Media | Reddy, D. | Springer Academic Publishers | Dordrecht | 2008 | ISBN 978-1-4020-9089-9 | | |
| 08-2 | | Modelling Nanomaterials and Nanosystems | Pyrz, R. and Rauhe, J.C. | Springer Academic Publishers | Dordrecht | 2009 | ISBN 978-1-4020-9556-6 | | |
| 08-3 | | Cellular, Molecular and Tissue Mechanics | Garikipati, K., Arruda, E.M. | Springer Academic Publishers | Dordrecht | 2010 | ISBN 978-90-481-3347-5 | | |
| 08-4 | | Variational Concepts with Applications to the Mechanics of Materials | Hackl, K. | Springer Academic Publishers | Dordrecht | 2010 | ISBN 978-90-481-9194-9 | | |
| 08-5 | | 150 Years of Vortex Dynamics | Aref, H. | Springer Academic Publishers | Dordrecht | 2010 | ISBN 978-90-481-8583-2 | | |
| 08-6 | | Progress in the Theory and Numerics of Configurational Mechanics | Steinmann, P. | Springer Academic Publishers | Dordrecht | 2009 | ISBN 978-90-481-3446-5 | | |
| 08-7 | | Rotating Stratified Turbulence and Turbulence in the Atmosphere and Oceans | Dritschel, D. | Springer | | 2010 | ISBN 978-94-007-0359-9 | | |
| 08-8 | | Multi Functional Material Structures and Systems | Dattaguru, B., Gopalakrishnan, S. and Aatre, V. K. | Springer Academic Publishers | Dordrecht | 2010 | ISBN 978-90-481-3770-1 | | |
| 09-1 | | Dynamic Fracture and Fragmentation | Ravi-Chandar, K., Vogler, T.J. | | | 2010 | International Journal of Fracture | | |
| 09-2 | | Emerging Trends in Rotor Dynamics | Gupta, K. | Springer | | 2011 | ISBN 978-94-007-0019-2 | | |

| Symposia | Total No. Papers | Symposia Name | Editors of Books (or Special Issue of Journals) | Publisher | City | Year | Name | Volume No. | Issue No. |
|---|---|---|---|---|---|---|---|---|---|
| 09-3 | | Recent Advances of Acoustic Waves in Solids | Wu, T.-T. and Ma, C.-C. | Springer | | 2010 | ISBN 978-90-481-9892-4 | | |
| 09-4 | | Laminar-Turbulent Transition | Schlatter, P., Henningson, D.S. | Springer Academic Publishers | Dordrecht | 2010 | ISBN 978-90-481-3722-0 | | |
| 09-5 | | Symposium on Vibration Analysis of Structures with Uncertainties | Belyaev, A.K. and Langley, R.S. | Springer | | 2011 | ISBN 978-94-007-0288-2 | | |
| 09-6 | | Wall-bounded Turbulent Flows on Rough Walls | Nickels, T.B. | Springer Academic Publishers | Dordrecht | 2010 | ISBN 978-90-481-9630-2 | | |
| 09-7 | | Multiscale Modelling of Fatigue, Damage and Fracture in Smart Materials Systems | Kuna, M., Ricoeur, A. | Springer | | 2011 | ISBN 978-90-481-9886-3 | | |
| 09-8 | | Mathematical Modeling and Physical Instances of Granular Flows | Goddard, J., Jenkins, J.T. and Giovine, P. | | | 2010 | AIP Conference Proceedings 1227 ISBN 978-0-7354-0772-5 | | |
| 10-1 | | Computational Aero-Acoustics for Aircraft Noise Prediction | Astley, J. and Gabard, G. | | | 2011 | Procedia IUTAM | 1 | |
| 10-2 | | Nonlinear Stochastic Dynamics and Control | Zhu, W.Q., Lin, Y.K., Cai, G. Q. | Springer | | 2011 | ISBN 978-94-007-0731-3 | | |
| 10-3 | | Dynamics Modeling and Interaction Control in Virtual and Real Environments | Stépán, G., Kovács, L. and Tóth, A. | Springer | | 2010 | ISBN 978-94-007-1642-1 | | |
| 10-4 | | Bluff Body Wakes and Vortex-Induced Vibrations | Leweke, T. and Williamson, C. | | | 8/ 2011 | Journal of Fluids and Structures | 27 | 5-6 |

| Symposia | Total No. Papers | Symposia Name | Editors of Books (or Special Issue of Journals) | Publisher | City | Year | Name | Volume No. | Issue No. |
|---|---|---|---|---|---|---|---|---|---|
| 10-5 | | Nonlinear Dynamics for Advanced Technologies and Engineering Design (NDATED) | Wiercigroch, M. and Rega, G. | Springer | | 2013 | ISBN 978-94-007-5742-4 | | |
| 10-6 | | Surface Effects in the Mechanics of Nanomaterials and Heterostructures | Cocks, A. and Wang, J. | Springer | | 2012 | ISBN 978-94-007-4910-8 | | |
| 10-7 | | Analysis and Simulation of Human Motion | Jonkers, I. and Vander Sloten, J. | published online | Leuven | 2010 | ISBN 978-94-6018-247-1 | | |
| 11-1 | | Mechanics of Liquid and Solid Foams | Kyriakides, S. and Kraynik, A. | | | 2012 | International Journal of Solids and Structures | 56 | 3 |
| 11-2 | | Linking Scales in Computations: From Microstructure to Macroscale Properties | Cazacu, O. | | | 2012 | Procedia IUTAM | 3 | |
| 11-3 | | Human Body Dynamics: From Multibody Systems to Biomechanics | McPhee, J., Kovecses, J. | | | 2011 | Procedia IUTAM | 2 | |
| 11-4 | | Full-field Measurements and Identification in Solid Mechanics | Hild, F. and Espinosa, H.D. | | | 2011 | Procedia IUTAM | 4 | |
| 11-5 | | Impact Biomechanics in Sport | Gilchrist, M., and Forero Rueda, M. | | | 2012 | Proceedings of the Institution of Mechanical Engineers, Part P: Journal of Sports Engineering and Technology | 226 | 3-4 |
| 11-6 | | Computer Models in Biomechanics: from Nano to Macro | Holzapfel, G.A. and Kuhl, E. | Springer | | 2013 | ISBN 978-94-007-5464-5 | | |

| Symposia | Total No. Papers | Symposia Name | Editors of Books (or Special Issue of Journals) | Publisher | City | Year | Name | Volume No. | Issue No. |
|---|---|---|---|---|---|---|---|---|---|
| 11-7 | | 50 Years of Chaos: Applied and Theoretical | Hikihara, T. | | | 2012 | Procedia IUTAM | 5 | |
| 11-8 | | Bluff Body Flows | Mittal, S. and Biswas, G. | | | 2013 | Journal of Fluids and Structures | 41 | |
| 12-1 | | Mobile Particulate Systems – Kinematics, Rheology and Complex Phenomena | Nott, P.R., Davis, R.H., Reeks, M., Saintillan, D. and Sundaresan, S. | | | 2013 | Physics of Fluids | 25 | 7 |
| 12-2 | | Advanced Materials Modelling for Structures | Altenbach, H. and Kruch, S. | Springer | | 2013 | ISBN 978-3-642-35167-9 | | |
| 12-3 | | From Mechanical to Biological Systems - an Integrated Approach | Kozlov, V.V. and Borisov, A.V. | | | | Regular and Chaotic Dynamics | 18 | 1-2 |
| 12-4 | | Waves in Fluids: Effects of Nonlinearity, Rotation, Stratification and Dissipation | Chashechkin, Y. and Dritschel, D. | | | 2013 | Procedia IUTAM | 8 | |
| 12-5 | | Multiscale Problems in Stochastic Mechanics | Proppe, C., Bourinet, J.-M. | | | 2013 | Procedia IUTAM | 6 | |
| 12-6 | | Fracture Phenomena in Nature and Technology | Bigoni, D.. Carini, A, Gei, M. and Salvadori, A. | | | 2013 | Int. Journal of Fracture | 184 | 1-2 |
| 12-7 | | Understanding Common Aspects of Extreme Events in Fluids | Bustamante, M., Newell, A.C., Kerr, R.M. and Tsubota, M. | | | 2013 | Procedia IUTAM | 9 | |
| 12-8 | | Topological Fluid Dynamics II | Moffatt, H.K., Bajer, K. and Kimura, Y. | | | 2013 | Procedia IUTAM | 7 | |
| 12-9 | | no proceedings | | | | | | | |

| Symposia | Total No. Papers | Symposia Name | Editors of Books (or Special Issue of Journals) | Publisher | City | Year | Name | Volume No. | Issue No. |
|---|---|---|---|---|---|---|---|---|---|
| 12-10 | 10 | Particle Methods in Fluid Mechanics | | | | | Procedia IUTAM | ? | |
| 12-11 | | no proceedings | | | | | | | |
| 13-1 | | Vortex Dynamics: Formation, Structure and Function | Fukumoto, Y. | | | 2014 | Fluid Dynamics Research | 46 | 3 |
| 13-2 | | Nonlinear Interfacial Wave Phenomena from the Micro- to the Macro-scale | Papageorgiou D., Smyrlis, Y., Vanden-Broeck J.-M. and Christodoulides, P. | | | 2014 | Procedia IUTAM | 11 | |
| 13-3 | | Recent Development of Experimental Techniques Under Impact Loading | Y. Li and H. Zhao | | | 2015 | International Journal of Impact Engineering | 79 | |
| 13-4 | | Materials and Interfaces under High Strain Rate and Large Deformation | Mercier, S. and Molinari, J.F. | | | 2015 | Mechanics of Materials | 80 | |
| 13-5 | | Multiscale Modeling and Uncertainty Quantification of Materials and Structures | Papadrakakis, M. and Stefanou, G. | Springer | | 2014 | ISBN 978-3-319-06330-0 | | |
| 13-6 | | The Dynamics of Extreme Events Influenced by Climate Change | Huang, N. | | | | Procedia IUTAM | 17 | |
| 14-1 | | Transition and Turbulence in the Flow through Deformable Tubes and Channels | Shankar, V. and Kumaran, V. | | | 2015 | Sadhana Acad. Proceedings in Engineering Sciences | 40 | 3 |
| 14-2 | | Mechanics of Soft Active Materials | Volokh, K. and Jabareen, M. | | | | Procedia IUTAM | 12 | |

| Symposia | Total No. Papers | Symposia Name | Editors of Books (or Special Issue of Journals) | Publisher | City | Year | Name | Volume No. | Issue No. |
|---|---|---|---|---|---|---|---|---|---|
| 14-5 | | Dynamical Analysis of Multibody Systems with Design Uncertainties | Hanss, M. | | | | Procedia IUTAM | 13 | |
| 14-6 | | no proceedings | | | | | | | |
| 14-7 | | Dynamics of Capsules, Vesicles and Cells in Flow | Barthes-Biesel, D., Blyth M., and Salsac, A.-V. | | | | Procedia IUTAM | 16 | |
| 14-10 | | Multiphase Flows with Phase Change: Challenges and Opportunities | Sahu, K. | | | | Procedia IUTAM | 15 | |

# Appendix 15
# General Assemblies

| Year | Month | City | Country |
|------|-------|------|---------|
| 1946 | September | Paris | France |
| 1948 | September | London | UK |
| 1950 | June | Pallanza | Italy |
| 1952 | August | Istanbul | Turkey |
| 1954 | July | Bruxelles | Belgium |
| 1956 | September | Bruxelles | Belgium |
| 1958 | October | Washington | USA |
| 1960 | September | Stresa | Italy |
| 1962 | September | Aachen | Germany |
| 1964 | September | Munich | Germany |
| 1966 | June | Vienna | Austria |
| 1968 | August | Stanford | USA |
| 1970 | August | Liège | Belgium |
| 1972 | August | Moscow | USSR |
| 1974 | September | Udine | Italy |
| 1976 | August | Delft | Netherlands |
| 1978 | September | Bad Herrenalb | Germany |
| 1980 | August | Toronto | Canada |
| 1982 | September | Cambridge | UK |
| 1984 | August | Lyngby | Denmark |
| 1986 | August | London | UK |
| 1988 | August | Grenoble | France |
| 1990 | September | Vienna | Austria |
| 1992 | August | Haifa | Israel |
| 1994 | August | Amsterdam | Netherlands |
| 1996 | August | Kyoto | Japan |
| 1998 | August | Stuttgart | Germany |
| 2000 | August | Chicago | USA |
| 2002 | August | Cambridge | UK |
| 2004 | August | Warsaw | Poland |
| 2006 | August | Providence | USA |
| 2008 | August | Adelaide | Australia |
| 2010 | July | Paris | France |
| 2012 | August | Beijing | China |
| 2014 | August | Lyngby | Denmark |
| 2016 | August | Montreal | Canada |

© The Author(s) 2016
P. Eberhard and S. Juhasz (eds.), *IUTAM*,
DOI 10.1007/978-3-319-31063-3

# Appendix 16
# General Assembly Members (1948–2014)

| Name | Country | Years | Total |
|------|---------|-------|-------|
| Abramson, H.N. | USA | 1987-90 | 4 |
| Achenbach, J.D. | USA | 1996-97, 09-14 | 8 |
| Ackeret, J. | Switzerland | 1948-79 | 32 |
| Acrivos, A. | USA | 1987-90, 04-14 | 15 |
| Adrian, R. | USA | 1998-01 | 4 |
| Afgan, N. | Yugoslavia | 1987-96 | 10 |
| Al-Athel, S. | Saudi Arabia | 1988-06 | 19 |
| Albring, W. | GDR | 1973 86 | 14 |
| Allen, D.N. de G. | UK | 1960-62 | 3 |
| Al-Suwaiyel, M.I. | Saudi Arabia | 2007-14 | 8 |
| Altenbach, J. | GDR | 1987-92 | 6 |
| Anchev, A. | Bulgaria | 1975-77 | 3 |
| Appa Rao, T.V.S.R. | India | 2000-01 | 2 |
| Aranha, J.A.P. | Brazil | 2007-08 | 2 |
| Aref, H. | USA | 1994-08 | 15 |
| Arinc, F. | Turkey | 1997-01 | 5 |
| Ariza, P. | Spain | 2014 | 1 |
| Arruda, F. | Brazil | 2009-12 | 4 |
| Astley, J. | New Zealand | 2002 | 1 |
| Aubry, N. | USA | 2007-14 | 8 |
| Baes, L. | Belgium | 1948-60 | 13 |
| Bai, Y.L. | China | 2001-14 | 14 |
| Baines, W.D. | Canada | 1964-70 | 7 |
| Baltov, A. | Bulgaria | 2004-10 | 7 |
| Bark, F. | Sweden | 1991-96 | 6 |
| Barthes-Biesel, D. | France | 1996-03 | 8 |
| Batchelor, G.K. | UK | 1958-99 | 42 |
| Battjes, J.A. | Netherlands | 1991-03 | 13 |
| Bautista, M. | Spain | 1984-88 | 5 |
| Becker, E. | FRG | 1966-84 | 19 |
| Béghin, M.H. | France | 1950 | 1 |
| Belhaq, M. | Marocco | 1998-03 | 6 |
| Belotserkovsky, O.M. | USSR | 1980-86 | 7 |
| Belytschko, T. | USA | 2001-06 | 6 |
| Benallal, A. | France | 2004-12 | 9 |
| Benjamin, T.B. | UK | 1982-87 | 6 |
| Berndt, S. | Sweden | 1966-90 | 25 |
| Besseling, J.F. | Netherlands | 1964-69, 80-91 | 18 |

© The Author(s) 2016
P. Eberhard and S. Juhasz (eds.), *IUTAM*,
DOI 10.1007/978-3-319-31063-3

| Name | Country | Years | Total |
|------|---------|-------|-------|
| Bevilacqua, L. | Brazil | 1984-11 | 28 |
| Bhagavantam, S. | India | 1968 | 1 |
| Bhatnagar, P.L. | India | 1969-72 | 4 |
| Bianchi, G. | Italy | 1976-01 | 26 |
| Biezeno, C.B. | Netherlands | 1948-63 | 16 |
| Bigoni, D. | Italy | 2013-14 | 2 |
| Bilimovic, A. | Yugoslavia | 1952-53 | 2 |
| Biron, A. | Canada | 1973-77 | 5 |
| Bishop, R.E.D. | UK | 1963-72 | 10 |
| Biswas, G. | India | 2003-14 | 12 |
| Bjorgum, O. | Norway | 1958-60 | 3 |
| Blaauwendraad, J. | Netherlands | 1981-86 | 6 |
| Blackwell, J.H. | Canada | 1971-72 | 2 |
| Bodner, S.R. | Israel | 1981-02, 04-12 | 31 |
| Boley, B.A. | USA | 1976-14 | 39 |
| Bonnet, M. | France | 2000-01 | 2 |
| Boström, A. | Sweden | 1995-07 | 13 |
| Bottaro, A. | Italy | 2013-14 | 2 |
| Boulanger, P. | Belgium | 2004-14 | 11 |
| Bourgois, R. | Belgium | 1996-01 | 6 |
| Brankov, G. | Bulgaria | 1968-87 | 20 |
| Brcic, V. | Yugoslavia | 1977-80 | 4 |
| Brilla, J. | CSSR | 1991-06 | 16 |
| Broer, L.J.F. | Netherlands | 1958-65 | 8 |
| Brun, E.A. | France | 1972-78 | 7 |
| Brunt, D. | UK | 1950-53 | 4 |
| Buckens, F. | Belgium | 1963-81 | 19 |
| Budiansky, B. | USA | 1972-75 | 4 |
| Burczynski, T. | Poland | 2012-14 | 3 |
| Burgers, J.M. | USA | 1948-80 | 33 |
| Cadambe, V. | India | 1952 | 1 |
| Calladine, C.R. | UK | 1991-02 | 12 |
| Calvo, R. | Spain | 1957-71 | 15 |
| Camotin, D. | Portugal | 2009-14 | 6 |
| Candel, S. | France | 2009-14 | 6 |
| Caquot, A. | France | 1948-65 | 18 |
| Carafoli, E. | Romania | 1954-83 | 30 |
| Cardon, A. | Belgium | 1996-03 | 8 |
| Cardou, A. | Canada | 1978-79 | 2 |
| Carlsson, J. | Sweden | 1984-90 | 7 |
| Carpenter, P.W. | UK | 1996-06 | 11 |

| Name | Country | Years | Total |
|------|---------|-------|-------|
| Carrier, G.F. | USA | 1964-69 | 6 |
| Cercignani, C. | Italy | 1984-06 | 23 |
| Cernuschi, F. | Argentina | 1959-69 | 11 |
| Chadwick, P. | UK | 1973-74 | 2 |
| Chandrasekaran, V.A. | India | 1994-97 | 4 |
| Chang, C.C. | China-Tapei | 2010-14 | 5 |
| Charnock, H. | UK | 1984-86 | 3 |
| Charru, S. | France | 2009-14 | 6 |
| Chazy, M.J. | France | 1951-53 | 3 |
| Chen, W.H. | China | 1998-04 | 7 |
| Chernousko, C.L. | Russia | 2006-14 | 9 |
| Chernyi, G.G. | USSR | 1984-10 | 27 |
| Cherry, T.M. | Australia | 1964-65 | 2 |
| Chiu, H.H. | China | 1990-97 | 8 |
| Choi, H.S. | Korea | 1994-97 | 4 |
| Chong, M. | Australia | 2012 | 1 |
| Chou, P.Y. | China | 1948-52, 78-81 | 9 |
| Christensen, R.M. | USA | 1990-91 | 2 |
| Christescu, N.D. | Romania | 1995-08 | 14 |
| Christianovic, S.A. | USSR | 1948-51 | 4 |
| Collins, I.F. | New Zealand | 1984-87, 93-97 | 9 |
| Collins, L. | USA | 2009-10 | 2 |
| Colonnetti, O. | Italy | 1948-67 | 20 |
| Coppens, A. | Belgium | 1951-57 | 7 |
| Cowper, G.R. | Canada | 1973-77 | 5 |
| Crandall, S. H. | USA | 1972-77,98-01 | 10 |
| Crocco, G.A. | Italy | 1948-70 | 23 |
| Crochet, M. | Belgium | 1992-95 | 4 |
| Crocker, M. | USA | 2000-01 | 2 |
| Cui, E. | China | 2003-06 | 4 |
| Cveticanin, L.V. | Serbia | 2004-05 | 2 |
| Dally, J.W. | USA | 1982-86 | 5 |
| Darwin, C. | UK | 1950-53 | 4 |
| Davis, E. | New Zealand | 2000 | 1 |
| Dawoud, R.H. | Egypt | 1978-92 | 15 |
| de Arantes e Oliveira, E.R. | Portugal | 1971-99 | 29 |
| de Azcarraga, L. | Spain | 1972-77 | 6 |
| de Borst, R. | Netherlands | 2004-12 | 9 |
| de Botton, G. | Israel | 2014 | 1 |
| de Pater, C. | Netherlands | 1968-73 | 6 |
| de Wilde, W.P. | Belgium | 1992-95 | 4 |

| Name | Country | Years | Total |
|---|---|---|---|
| Dechaene, R. | Belgium | 1992-95 | 4 |
| Denier, J.P. | Australia, New Zealand | 2006-11, 2014 | 7 |
| Deshpande, S.M. | India | 2003-08 | 6 |
| Dias, F. | France | 2004-14 | 11 |
| Dick, E. | Belgium | 2004-14 | 11 |
| Del Pedro, M. | Switzerland | 1977-83 | 7 |
| Dikmen, M. | Turkey | 1977-79, 83-89 | 10 |
| Dijksman, J.F. | Netherlands | 1993-03 | 11 |
| Djordjevic, V. | Yugoslavia | 1981 | 1 |
| Do Toit, C.G. | South Africa | 1998-05 | 8 |
| Dolapchiev, B. | Bulgaria | 1968-72 | 5 |
| Dorado, R. | Spain | 1998 | 1 |
| Dost, S. | Canada | 1993-98, 2014 | 7 |
| Dowell, E.H. | USA | 1996-00 | 5 |
| Droogge, G. | Sweden | 1988-94 | 7 |
| Drucker, D.C. | USA | 1964-00 | 37 |
| Dryden, H.L. | USA | 1948-65 | 18 |
| Dual, J. | Switzerland | 2002-14 | 13 |
| Dubuc, J. | Canada | 1966-70 | 5 |
| Duncan, B.J. | UK | 1954-59 | 6 |
| Dvorak, R. | CSSR | 1996-06 | 11 |
| Eberhard, P. | Germany | 2005-14 | 10 |
| Ellyn, F. | Canada | 2000-01 | 2 |
| Emmons, H.W. | USA | 1950-51 | 2 |
| Ene, H. | Romania | 2009-14 | 6 |
| Engelbrecht, J. | Estonia | 1992-08 | 17 |
| Engelund, F. | Denmark | 1971-82 | 12 |
| Erim, K. | Turkey | 1948-51 | 4 |
| Espedal, M.S. | Norway | 1987-88 | 2 |
| Favre, H. | Switzerland | 1948-66 | 19 |
| Federhofer, K. | Austria | 1951-59 | 9 |
| Ferarri, C. | Italy | 1970-75 | 6 |
| Finzi, B. | Italy | 1966-69 | 4 |
| Fiszdon, W. | Poland | 1971-03 | 33 |
| Flavin, J. | Ireland | 1992-95 | 4 |
| Fleck, N.A. | UK | 2007-14 | 8 |
| Floryan, M. | Canada | 2009-14 | 6 |
| Förste, J. | GDR | 1973-77 | 5 |
| Fraeijs de Veubeke, B. | Belgium | 1971-75 | 5 |
| Franzoni, L.P. | USA | 2011-14 | 4 |

| Name | Country | Years | Total |
|---|---|---|---|
| Freire, A. | Brazil | 2013-14 | 2 |
| Frenkiel, F.N. | USA | 1954-86 | 33 |
| Freund, L.B. | USA | 1987-14 | 28 |
| Frolov, K.V. | USSR | 1987-00 | 14 |
| Gabbert, O. | Germany | 1994-04 | 11 |
| Gadambe, V. | India | 1953-56 | 4 |
| Galletto, D. | Italy | 1976-93 | 18 |
| Gauthier, l. | France | 1966-78 | 13 |
| Geers, M. | Netherlands | 2013-14 | 2 |
| Geleji, S. | Hungary | 1963-67 | 5 |
| Germain, P. | France | 1966-08 | 43 |
| Gersten, K. | FRG | 1987-93 | 7 |
| Gilchrist, M. | Ireland | 2009-14 | 6 |
| Gjevik, B. | Norway | 1989-06 | 18 |
| Glockner, P.G. | Canada | 1971-75 | 5 |
| Goldhirsch, I. | Israel | 2002-09 | 8 |
| Goldstein, S. | USA | 1948-87 | 40 |
| Gopalakrishna, S. | India | 2011 | 1 |
| Goriacheva, I.G. | Russia | 2011-14 | 4 |
| Görtler, H. | FRG | 1952-81 | 30 |
| Gough, H.J. | UK | 1950-53 | 4 |
| Gradowczyk, M. | Argentina | 1970-91 | 22 |
| Gramel, R. | FRG | 1950-63 | 14 |
| Green, H.S. | Australia | 1966-68 | 3 |
| Grue, J. | Norway | 2007-14 | 8 |
| Guiraud, J.P. | France | 1979-83 | 5 |
| Günther, H. | GDR | 1987-90 | 4 |
| Gupta, A.K. | India | 1989-91 | 3 |
| Gupta, N. | India | 2000-12 | 13 |
| Gutkowski, W. | Poland | 1987-11 | 25 |
| Guz, A.N. | Ukraine | 1995-14 | 20 |
| Hacar, B. | CSSR | 1958-59 | 2 |
| Hall, A.A. | UK | 1950-62 | 13 |
| Hansen, E. | Denmark | 1984-90 | 7 |
| Hansen, J. | Canada | 1999-08 | 10 |
| Hashin, Z. | Israel | 1976-94 | 19 |
| Haus, F. | Belgium | 1963- 70 | 8 |
| Hayes, M.A. | Ireland | 1984-87, 93-08 | 20 |
| He, Y.S. | China | 1990-02 | 13 |
| Heijden, A.M.A.v. | Netherlands | 1987-90 | 4 |
| van Heijst, G.J.F. | Netherlands | 2012-14 | 3 |

| Name | Country | Years | Total |
|---|---|---|---|
| Hennig, K. | GDR | 1978-86 | 9 |
| Henningson, D. | Sweden | 2005-14 | 10 |
| Herakovich, C.T. | USA | 1998-14 | 17 |
| Hetenyi, M. | USA | 1957-63 | 7 |
| Hinze, J.O. | Netherlands | 1966-73 | 8 |
| Hirsch, P. | UK | 1984-86 | 3 |
| Hoa Thinh, N. | Vietnam | 2009-12 | 4 |
| Hodge, P.G., jr. | USA | 1982-08 | 27 |
| Hoff, N.J. | USA | 1950-96 | 47 |
| Horak, Z. | CSSR | 1950-57 | 7 |
| Hornung, H. | FRG | 1982-86 | 5 |
| Hostinsky, B. | CSSR | 1948-52 | 5 |
| Hu, H. | China | 2012-14 | 3 |
| Huber, M.T. | Poland | 1948-50 | 3 |
| Hudnett, P.F. | Ireland | 1996-99 | 4 |
| Hult, J. | Sweden | 1970-12 | 43 |
| Hunsaker, J.C. | USA | 1948-53 | 6 |
| Hwang, G.J. | China | 1988-97 | 10 |
| Hwang, K.C. | China | 1990-00 | 11 |
| Idelsohn, S.R. | Argentina | 1992-11 | 20 |
| Ikeda, S. | Japan | 1954-59 | 6 |
| Ilyushin, A.A. | USSR | 1954-75 | 22 |
| Imai, I. | Japan | 1960-99 | 40 |
| Inan, E. | Turkey | 1992-94 | 3 |
| Inoue, T. | Japan | 1993-99 | 7 |
| Iooss, G. | France | 1984-95 | 12 |
| Ippen, A.T. | USA | 1952-56 | 5 |
| Ishlinsky, A.Y. | USSR | 1976-02 | 27 |
| Ismail, M.A. | Egypt | 1994-14 | 21 |
| Ito, M. | Japan | 1996-99 | 4 |
| Iyengar | India | 1989-91 | 3 |
| Jaiani, G. | Georgia | 2000-14 | 15 |
| Janssens, P. | Belgium | 1971-91 | 21 |
| Jaric, J. | Yugoslavia | 1982-86 | 5 |
| Jecic, S. | Yugoslavia | 1974-80 | 7 |
| Jerath, G.C. | India | 1950 | 1 |
| Jerie, J. | CSSR | 1964-83 | 20 |
| Johansen, K.W. | Denmark | 1960-61 | 2 |
| Jones, N. | UK | 2001-02 | 2 |
| Joseph, D. | USA | 1991 | 1 |
| Kaliszky, S. | Hungary | 1984-11 | 28 |

| Name | Country | Years | Total |
|------|---------|-------|-------|
| Kambe, T. | Japan | 1996-12 | 17 |
| Kant, T. | India | 2000-02, 09-10 | 5 |
| Karev, V. | Russia | 2011-12 | 2 |
| Karihaloo, B.L. | Australia | 1988-01, 03-10,12-14 | 25 |
| Katsikadelis, J.T. | Greece | 2007-11 | 5 |
| Kawata, K. | Japan | 1983-95 | 13 |
| Keffer, J.F. | Canada | 1971-72 | 2 |
| Keller, J.B. | USA | 1982-86 | 5 |
| Kerwin, L. | Canada | 1980 | 1 |
| Kestens, J. | Belgium | 1984-91 | 8 |
| Keunings, R. | Belgium | 1996-03 | 8 |
| Kienzler, R. | Germany | 2009-14 | 6 |
| Kishimoto, K. | Japan | 2011-14 | 4 |
| Kitagawa, H. | Japan | 2000-03 | 4 |
| Klimov, D. | Russia | 2001-05 | 5 |
| Kluwick, A. | Austria | 1991-14 | 24 |
| Knauss, W. | Austria | 2002-05 | 4 |
| Kobayashi, T. | Japan | 2000-10 | 11 |
| Koiter, W.T. | Netherlands | 1952-63, 1969-97 | 11 |
| Kok, S. | South Africa | 2009-14 | 6 |
| Kolarov, D. | Bulgaria | 1978-87 | 10 |
| Kolmogoroff, A.N. | USSR | 1948-51 | 4 |
| Kondo, D. | France | 2013-14 | 2 |
| Koning, C. | Netherlands | 1950-51 | 2 |
| Kouhia, R. | Finland | 2013-14 | 2 |
| Kounadis, A.N. | Greece | 1987-06 | 20 |
| Kozesnik, J. | CSSR | 1960-83 | 24 |
| Krasnikovs, A. | Latvia | 2012-13 | 2 |
| Krause, E. | FRG | 1900-00 | 11 |
| Krupka, V. | CSSR | 1984-90 | 7 |
| Ku, Y.H. | USA | 1948-03 | 56 |
| Küchemann, D. | UK | 1961-70 | 10 |
| Kuhelj, A. | Yugoslavia | 1954-73 | 20 |
| Kuhn, G. | Germany | 1988-00 | 13 |
| Kuyumdzhiev, H. | Bulgaria | 2001-03 | 3 |
| Kwak, B.M. | Korea | 1998-99 | 2 |
| Kyriadikes, S. | USA | 2010-14 | 5 |
| Lafita, F. | Spain | 1950-51 | 2 |
| Larsen, P.S. | Denmark | 1996-97 | 2 |
| Lavrentiev, M.A. | USSR | 1978-79 | 2 |
| Leal, G. | USA | 1992-97, 01-08 | 14 |

| Name | Country | Years | Total |
|---|---|---|---|
| Lee, C.M. | Korea | 1990-93 | 4 |
| Lee, E.H. | USA | 1969-75 | 7 |
| Legendre, R. | France | 1966-83 | 18 |
| Leibovich, S. | USA | 1991-93 | 3 |
| Leipholz, H. | Canada | 1973-79 | 7 |
| Lemaitre, M.G. | Belgium | 1958-62 | 5 |
| Lespinard, G. | France | 1984-91 | 8 |
| Leung, A.Y.T. | Hongkong | 2011-14 | 4 |
| Levin, V. | Russia | 2013-2014 | 2 |
| Li, J. | China | 2007-14 | 8 |
| Lighthill, M.J. | UK | 1954-57, 64-97 | 38 |
| Lin, T.C. | China | 1980-89 | 10 |
| Lippmann, H. | FRG | 1975-87 | 13 |
| Lugner, P. | Austria | 2001 | 1 |
| Lunc, M. | Poland | 1954-69 | 16 |
| Lund, F. | Chile | 1996-14 | 19 |
| Lundberg, B. | Sweden | 1991-11 | 21 |
| Lundstrom, S. | Sweden | 2009-14 | 6 |
| Luthander, S. | Sweden | 1961-65 | 5 |
| Ma, C.-C. | China-Tapei | 2005-10 | 6 |
| Määttänen, M. | Finland | 1995-08 | 14 |
| Madsen, P.A. | Denmark | 1998-02 | 5 |
| Mahony, J.J. | Australia | 1975-80 | 6 |
| Mahrenholtz, O. | Germany | 1993 | 1 |
| Maier, G. | Italy | 1976-83, 00-14 | 23 |
| Majumdar, B.C. | India | 1994-97 | 4 |
| Makris, N. | Greece | 2012-14 | 3 |
| Maksimovic, S. | Serbia | 2009-12 | 4 |
| Mallik, A.K. | India | 1997-99 | 3 |
| Mansa, J.L. | Denmark | 1950-59 | 10 |
| Mansfield, E.H. | UK | 1975-78 | 4 |
| Martin, J.B. | South Africa | 1994-95 | 2 |
| Martin, M.H. | USA | 1952-53 | 2 |
| Martins, J. | Portugal | 2000-08 | 9 |
| Marusic, Y. | Australia | 2012-14 | 3 |
| Matsumoto, Y. | Japan | 2011-14 | 4 |
| Maunder, L. | UK | 1970-78 | 9 |
| McCarthy, M.F. | Ireland | 1990-91 | 2 |
| McVerry, G. | New Zealand | 1988-93 | 6 |
| Meijers, P. | Netherlands | 1974-80 | 7 |
| Melan, E. | Austria | 1960-62 | 3 |

| Name | Country | Years | Total |
|---|---|---|---|
| Mertens, R. | Belgium | 1978-87 | 10 |
| Mettler, E. | FRG | 1964-74 | 11 |
| Miehe, C. | Germany | 2001-08 | 8 |
| Mikhailov, G.K. | USSR | 1976-10 | 35 |
| Mikkola, M.J. | USSR | 1988-00 | 13 |
| Miller, K. | UK | 1984-92 | 9 |
| Miloh, T. | Israel | 1995-01 | 7 |
| Mindlin, R.D. | USA | 1950-56 | 7 |
| Mittal, S. | India | 2011-14 | 4 |
| Moffatt, H.K. | UK | 1979-83, 88-14 | 32 |
| Mogami, T. | Japan | 1952-53 | 2 |
| Molinari, A. | France | 2009-14 | 6 |
| Monkewitz, P.A. | Switzerland | 1994-14 | 21 |
| Moratilla, A. | Spain | 1999-13 | 15 |
| Moreau, R. | France | 1992-01 | 10 |
| Moreno Labata, G. | Spain | 1996-97 | 2 |
| Morozov, N. | Russia | 2001-14 | 14 |
| Morro, A. | Italy | 2007-14 | 8 |
| Morsy, H.A. | Egypt | 1993 | 1 |
| Munjal, M.N. | India | 1997-02 | 6 |
| Muskhelishvili, N.I. | USSR | 1969-75 | 7 |
| Naghdi, P.M. | USA | 1978-86 | 9 |
| Nakagawa, H. | Japan | 1992-95 | 4 |
| Nakanishi, F. | Japan | 1952-53 | 2 |
| Nakra, B.C. | India | 1992-93 | 2 |
| Napolitano, L.G. | Italy | 1976-83 | 8 |
| Narasimha, R. | India | 1982-87,92-98,01-03,09-14 | 22 |
| Nath, Y. | India | 1988-93 | 6 |
| Neale, K.W. | Canada | 1978-86 | 9 |
| Nemec, J. | CSSR | 1984-89 | 6 |
| Newmark, N.M. | USA | 1952-56 | 5 |
| Nguyen, Van Dao | Vietnam | 1990-2006 | 17 |
| Nicholl, C.I.H. | Canada | 1964-65 | 2 |
| Nichols, R.W. | UK | 1978-80 | 3 |
| Nicolai, E.L. | USSR | 1948-50 | 3 |
| Nielsen, J. | Denmark | 1948-51 | 4 |
| Nigam, N.C. | India | 1982-87 | 6 |
| Nigam, S.D. | India | 1975-77 | 3 |
| Nikolskii, A.A. | USSR | 1954-75 | 22 |
| Niordson, C. | Denmark | 2012-14 | 3 |

| Name | Country | Years | Total |
|---|---|---|---|
| Niordson, F.I. | Denmark | 1962-08 | 47 |
| Nishimura, N. | Japan | 2011-14 | 4 |
| Norris, A. | USA | 2008 | 1 |
| Nowacki, W. | Poland | 1954, 69-78 | 11 |
| Nunez, A. | Spain | 1954-56 | 3 |
| O`Donoghue, P. | Ireland | 2000-08 | 9 |
| Oberlack, M. | Germany | 2009-14 | 6 |
| Obraztsov, I.M. | USSR | 1980-00 | 21 |
| Oden, J.T. | USA | 1991-01 | 11 |
| Odqvist, F.K.G. | Sweden | 1950-83 | 34 |
| Ohashi, H. | Japan | 1996-99 | 4 |
| Okrouhlik, M. | Czech Rep. | 2007-14 | 8 |
| Okubo, H. | Japan | 1960-62 | 3 |
| Olhoff, N. | Denmark | 1991-14 | 24 |
| Oliver, D. | Spain | 1973-75 | 3 |
| Olsen, R.G. | Norway | 1954-56 | 3 |
| Olszak, W. | Poland | 1954-79 | 26 |
| Oosterveld, M.W.C. | Netherlands | 1974-76 | 3 |
| Osuna, D.O. | Spain | 1972 | 1 |
| Owen, P.R. | UK | 1971-78 | 8 |
| Paavola, J. | Finland | 2001-14 | 14 |
| Pacejka, H.B. | Netherlands | 1978-88 | 11 |
| Pak, C.H. | Korea | 1994-97 | 4 |
| Palm, E. | Norway | 1966-73 | 8 |
| Parkus, H. | Austria | 1963-81 | 19 |
| Parushev, P. | Bulgaria | 1988-93 | 6 |
| Peake, N. | UK | 2003-14 | 12 |
| Pearson, J.R.A. | UK | 1974-01 | 28 |
| Pedersen, P.T. | Denmark | 1987, 89-97 | 10 |
| Pedley, T.J. | UK | 1993-14 | 22 |
| Peres, J. | France | 1948-60 | 13 |
| Perez del Puerto, G. | Spain | 1976-83 | 8 |
| Perez-Marin, A. | Spain | 1951-71 | 21 |
| Pestel, E. | FRG | 1970-77 | 8 |
| Petryk, H. | Poland | 2012-14 | 3 |
| Phan-Thien, N. | Australia | 1996-05 | 10 |
| Phillip, J.R. | Australia | 1984-86 | 3 |
| Pilipenko, V.V. | Ukraine | 1995-1997 | 3 |
| Pin Tong | Hong Kong | 1997 | 1 |
| Podio-Guidugli, P. | Italy | 1994-12 | 19 |
| Polizzotto, C. | Italy | 1987-97 | 11 |

| Name | Country | Years | Total |
|------|---------|-------|-------|
| Popoff, K. | Bulgaria | 1948-66 | 19 |
| Prager, W. | USA | 1957-71 | 15 |
| Prata, A. | Brazil | 2012-14 | 3 |
| Prathap, G. | India | 1992-93 | 2 |
| Prentis, J.M. | UK | 1979-83 | 5 |
| Price, W.G. | UK | 1988-90, 94-95 | 5 |
| Quian, L. | China | 1987-89 | 3 |
| Quintana Gonzalez, J.M. | Spain | 1989-95 | 7 |
| Rabcitnov, Yu. N. | USSR | 1976-79 | 4 |
| Radev, S. | Bulgaria | 1988-00,2011-14 | 17 |
| Radok, J.R.M. | Australia | 1969-74 | 6 |
| Ramanurti, V. | India | 1994-97 | 4 |
| Ramachandran, A. | India | 1973-74 | 2 |
| Ramberg, W. | USA | 1957-63 | 7 |
| Ramos, E. | Mexico | 2009-14 | 6 |
| Ranta, M.A. | Finland | 1971-94 | 24 |
| Reddy, B.D. | South Africa | 1996-97 | 2 |
| Rega, G. | Italy | 2009-14 | 6 |
| Reidjanovic, M. | Yugoslavia | 1981-83 | 3 |
| Reiner, M. | Israel | 1952-75 | 24 |
| Reuss, E. | Hungary | 1954-62 | 9 |
| Rimrott, F.P.J. | Canada | 1964-02 | 39 |
| Rivlin, R.S. | USA | 1974-83 | 10 |
| Rocha, M. | Portugal | 1968-70 | 3 |
| Rodriguez, A.M. | Spain | 1952-53 | 2 |
| Roseau, M. | France | 1984-96 | 13 |
| Rott, N. | Switzerland | 1967-82 | 16 |
| Roy, M. | France | 1950-85 | 36 |
| Rozvany, G. | Hungary | 1997-01 | 5 |
| Rubin, M.B. | Israel | 2003-14 | 12 |
| Rubinowicz, W. | Poland | 1948-74 | 27 |
| Ruzec, D. | Yugoslavia | 1987-03 | 17 |
| Ryhming, I.L. | Switzerland | 1984-93 | 10 |
| Saje, M. | Slowenia | 1994-01 | 8 |
| Salencon, J. | France | 1990-03 | 14 |
| Salupere, A. | Estonia | 2009-14 | 6 |
| Salvadori, M.G. | USA | 1950-51 | 2 |
| Salyi, I. | Hungary | 1968-72 | 5 |
| Sano, O. | Japan | 2013-14 | 2 |
| Sauer, R. | FRG | 1960-65 | 6 |
| Savage, S.B. | Canada | 1987-13 | 27 |

| Name | Country | Years | Total |
|---|---|---|---|
| Savic, P. | Canada | 1964-69 | 6 |
| Sawaragi, Y. | Japan | 1954-59 | 6 |
| Sawczuk, A. | Poland | 1979-83 | 5 |
| Sayir, M. | Switzerland | 1983-01 | 19 |
| Schiehlen, W. | Germany | 1978-14 | 37 |
| Schmidt, G. | GDR | 1973-86 | 14 |
| Schrefler, B.A. | Italy | 2012-14 | 3 |
| Schröder, W. | Germany | 2001-08 | 8 |
| Sedov, L.I. | USSR | 1954-98 | 45 |
| Seeger, R.J. | USA | 1950-51 | 2 |
| Seifried, R. | Germany | 2013-14 | 2 |
| Sengupta, S.R. | India | 1951-56 | 6 |
| Sestini, G. | Italy | 1973-75 | 3 |
| Sharma, V.D. | India | 2013-14 | 2 |
| Sharp, R.S. | UK | 1989-00 | 12 |
| Shrivastava, S. | Canada | 2006-14 | 9 |
| Singh, D. | India | 2009-10 | 2 |
| Skerget, I. | Slovenia | 2002-14 | 13 |
| Sladek, J. | Slovakia | 2007-13 | 7 |
| Sloan, S. | Australia | 2009-14 | 6 |
| Smetana, M.J. | France | 1960-62 | 3 |
| Smyrlis, Y.S. | Cyprus | 2012-14 | 3 |
| Sokolovskii, V.V. | USSR | 1954-75 | 22 |
| Solberg, H. | Norway | 1948-53 | 6 |
| Sobrero, L. | Italy | 1971-77 | 7 |
| Sorensen, J.N. | Denmark | 2003-14 | 12 |
| Southwell, R.V. | UK | 1948-70 | 23 |
| Spasic, D. | Serbia | 2013-14 | 2 |
| Squire, H.B. | UK | 1958-59 | 2 |
| Stahle, P. | Sweden | 2012-14 | 3 |
| Stenberg, R. | Finland | 2009-12 | 4 |
| Stenij, S.E. | Finland | 1952-70 | 19 |
| Stepan, G. | Hungary | 2012-14 | 3 |
| Storakers, B. | Sweden | 1997-04 | 8 |
| Stuart, J.T. | UK | 1984-92 | 9 |
| Suhubi, E. | Turkey | 1995-14 | 20 |
| Sukhatme, S.P. | India | 1978-81, 97-01 | 9 |
| Sumarac, G.M. | Serbia | 2006-08 | 3 |
| Sun, F.T. | China | 1980-86 | 7 |
| Sung, H.J. | Korea | 2012-14 | 3 |
| Suo, Z. | USA | 2007-14 | 8 |

| Name | Country | Years | Total |
|------|---------|-------|-------|
| Supino, G. | Italy | 1970-72 | 3 |
| Suquet, P. | France | 1996-03 | 8 |
| Sverdrup, H.U. | Norway | 1948-51 | 4 |
| Swanson, S.R. | Canada | 1980-91 | 12 |
| Szabo, J. | Hungary | 1973-83 | 11 |
| Szczepinski, W. | Poland | 1981-90 | 10 |
| Szefer, G. | Poland | 2001-11 | 11 |
| Tabarrok, B. | Canada | 1980-98 | 19 |
| Tamaki, F. | Japan | 1954-59 | 6 |
| Tamuzs, V. | Latvia | 1992-11 | 20 |
| Tani, I. | Japan | 1960-86 | 27 |
| Tanner, R.I. | Australia | 1981-83, 96-05 | 13 |
| Taplin, D.M.R. | Canada | 1981-83 | 3 |
| Tatsumi, T. | Japan | 1989-14 | 26 |
| Taylor, G.I. | UK | 1948-74 | 27 |
| Temple, G. | UK | 1950-90 | 41 |
| Terzioglu, N. | Turkey | 1969 | 1 |
| Tezel, A. | Turkey | 1980-82, 90-91 | 5 |
| Theocaris, P. | Greece | 1980-86 | 7 |
| Thess, A. | Germany | 2003-14 | 12 |
| Thien Khiem, N. | Vietnam | 2013-14 | 2 |
| Thiry, M.R. | France | 1958-65 | 8 |
| Timoshenko, S.P. | USA | 1948-70 | 23 |
| Tjøtta, S. | Norway | 1974-86 | 13 |
| Tomotika, S. | Japan | 1952-53,63-64 | 4 |
| Tong, P. | Hongkong | 1999-02 | 4 |
| Tuck, E.O. | Australia | 2006-08 | 3 |
| Turkalj, G. | Croatia | 2006-14 | 9 |
| Tvergaard, V. | Denmark | 2012-14 | 3 |
| Uetani, A. | Japan | 2004-05, 07-10 | 6 |
| Usher, S. | New Zealand | 1998-99 | 2 |
| van Steenhoven, A.A. | Netherlands | 2004-11 | 8 |
| Valluri, S.R. | India | 1969 | 1 |
| van Campen, D.H. | Netherlands | 1991-14 | 24 |
| van den Dungen, F.H. | Belgium | 1948-65 | 18 |
| Van Diep, N. | Vietnam | 2007-08 | 2 |
| van Eepoel, P. | Belgium | 1966-70 | 5 |
| van Moerbeke, P. | Belgium | 1988-90 | 3 |
| Vandepitte, D. | Belgium | 2002-14 | 13 |
| Vatta, F. | Italy | 1995-08 | 14 |
| Vekua, I.N. | USSR | 1976 | 1 |

| Name | Country | Years | Total |
|---|---|---|---|
| Villaggio, P. | Italy | 1984-86 | 3 |
| Villat, M.H. | France | 1952-65 | 14 |
| Viswanath, P. | India | 2000-01 | 2 |
| von Karman, T. | USA | 1948-60 | 13 |
| von Mises, R. | USA | 1948-52 | 5 |
| Vossers, O. | Netherlands | 1977-86 | 10 |
| Wagner, S. | Germany | 1994-02 | 9 |
| Walther, A. | FRG | 1952-65 | 14 |
| Wang, W.C. | China-Taipeh | 2005-14 | 10 |
| Wang, R. | China | 1989-00 | 12 |
| Washizu, K. | Japan | 1978-81 | 4 |
| Watanabe, E. | Japan | 2000-10 | 11 |
| Weaver, D.S. | Canada | 1999-05 | 7 |
| Weibull, W. | Sweden | 1948-78 | 31 |
| Weir, G. | New Zealand | 2001,03-13 | 12 |
| Wijngaarden, L. van | Netherlands | 1978-14 | 37 |
| Willis, J.R. | UK | 1989-00 | 12 |
| Wilson, W.B. | New Zealand | 1980-83 | 4 |
| Wittrick, W.H. | UK | 1979-86 | 8 |
| Yagawa, G. | Japan | 1996-99 | 4 |
| Yajnik, K.S. | India | 1978-81 | 4 |
| Yamamoto, Y. | Japan | 1978-95 | 18 |
| Yang, W. | China | 2001-14 | 14 |
| Yeh, C.S. | China | 1998-04 | 7 |
| Yokobori, T. | Japan | 1984-97 | 14 |
| Yoo, J.Y. | Korea | 2001-11 | 11 |
| Yoshiki, M. | Japan | 1965-77 | 13 |
| Yosibash, Z. | Israel | 2012-13 | 2 |
| Youm, Y. | Korea | 1990-93 | 4 |
| Yu, T.X. | Hongkong | 2003-10 | 8 |
| Yul-Yoh, J. | Korea | 2000 | 1 |
| Zaleski, S. | France | 2000-08 | 9 |
| Zaoui, A. | France | 2000-08 | 9 |
| Zaric, Z. | Yugoslavia | 1980-85 | 6 |
| Zheng, Z. | China | 1988-11 | 24 |
| Zhou, P.Y. | China | 1982-93 | 12 |
| Ziegler, F. | Austria | 1982-99, 04-08 | 23 |
| Ziegler, H. | Switzerland | 1950-76 | 27 |
| Zienkiewicz, O.C. | UK | 1988-90 | 3 |
| Zorski, H. | Poland | 1990-00 | 11 |
| Zu, J.W. | Canada | 2004-14 | 11 |

# Appendix 17
# Symposia Panels

Solids

| | chair | | | | |
|---|---|---|---|---|---|
| 1977-80 | Germain | Crandall | Koiter | Lippmann | |
| 1981-84 | Koiter | Crandall | Germain | Lippmann | |
| 1985-88 | Crandall | Maier | Germain | Lippmann | Zyczkowski |
| 1989-92 | Crandall | Maier | Achenbach | Salencon | Zyczkowski |
| 1993-96 | Salencon | Maier | Achenbach | Sobczyk | Storakers |
| 1997-00 | Salencon | Maier | Achenbach | Sobczyk | Storakers |
| 2001-04 | Achenbach | Willis | Ehlers | Tvergaard | Chernousko |
| 2005-06 | Achenbach | Willis | Ehlers | Tvergaard | Chernousko |
| 2007-08 | | | | | Stepan |
| 2009-10 | Tvergaard | Fleck | Ehlers | Gao | |
| 2011-12 | | | | | Stepan |
| 2013-14 | Fleck | Corigliano | Leblond | Gao | |
| 2015-16 | | | | | Lu |

Fluids

| | chair | | | | |
|---|---|---|---|---|---|
| 1977-80 | Lighthill | Fiszdon | Frenkiel | Sedov | |
| 1981-84 | Lighthill | Fiszdon | Frenkiel | Sedov | |
| 1985-88 | Moffatt | Acrivos | Guirand | Imai | Narashima |
| 1989-92 | Acrivos | Moffatt | Gersten | Imai | Narashima |
| 1993-96 | Acrivos | Moffatt | Gersten | Tatsumi | Narashima |
| 1997-00 | Acrivos | Huerre | Krause | Tatsumi | Narashima |
| 2001-04 | Huerre | Leal | Krause | Kambe | Peregrine |
| 2005-06 | Huerre | Leal | Hennigson | Kambe | Peregrine |
| 2007-08 | | | | Sreenivasan | Stiasnie |
| 2009-10 | Leal | Peake | | | |
| 2011-12 | | | Choi | Sreenivasan | Stiasnie |
| 2013-14 | Leal | Peake | | | |
| 2015-16 | | | Choi | Govindarajan | Lohse |

© The Author(s) 2016
P. Eberhard and S. Juhasz (eds.), *IUTAM*,
DOI 10.1007/978-3-319-31063-3

# Appendix 18
# Number of General Assembly Representatives by Countries

| 1 | Austria | Mexico |
|---|---|---|
| | Bulgaria | New Zealand |
| | Chile | Norway |
| | Croatia | Portugal |
| | Cyprus | Romania |
| | Czech Republic | Saudi Arabia |
| | Egypt | Serbia |
| | Estonia | Slovenia |
| | Georgia | South Africa |
| | Greece | Spain |
| | Hongkong | Turkey |
| | Hungary | Ukraine |
| | Ireland | Vietnam |
| | Korea | |

| 2 | Australia | Israel |
|---|---|---|
| | Brazil | Netherlands |
| | China/Taipei | Poland |
| | Denmark | Switzerland |
| | Finland | |

| 3 | Belgium | Sweden |
|---|---|---|
| | India | |

| 4 | Canada | Italy |
|---|---|---|
| | China | Japan |
| | France | Russia |
| | Germany | UK |

| 5 | USA | |
|---|---|---|

| suspended by decision of the GA | Argentina | Morocco |
|---|---|---|
| | Latvia | Slovakia |

data as of 2015, several changes in the number of representatives occurred over the years

© The Author(s) 2016
P. Eberhard and S. Juhasz (eds.), *IUTAM*,
DOI 10.1007/978-3-319-31063-3

# Appendix 19
# Adhering Organizations

*Argentina (1959–2012)*
Sociedad Argentina de Mecanica Teorica y Applicada Facultad de Ingenieria,

*Australia (1964)*
The Australian National Committee for Mechanical and Engineering Sciences of the Australian Academy of Science

*Austria (1951)*
Austrian National Committee for Theoretical and Applied Mechanics of the Austrian Academy of Sciences

*Belgium (1949)*
The National Committee for Theoretical and Applied Mechanics of the Royal Academies for Science and Arts of Belgium

*Brazil (1982)*
Associação Brasileira de Engenharia e Ciências Mecânicas—ABCM

*Bulgaria (1969)*
Bulgarian National Committee on Theoretical and Applied Mechanics of the Bulgarian Academy of Sciences

*Canada (1963)*
The National Research Council of Canada

*Chile (1996)*
The Chile National Committee on Theoretical and Applied Mechanics Academia Chilena de Ciencias

*China (1980)*
The Chinese Society of Theoretical and Applied Mechanics (CSTAM)

*China-Hong Kong (1996)*
The Hong Kong Society of Theoretical and Applied Mechanics (HKSTAM)

*China-Taipei (1980)*
The Society of Theoretical and Applied Mechanics

© The Author(s) 2016
P. Eberhard and S. Juhasz (eds.), *IUTAM*,
DOI 10.1007/978-3-319-31063-3

*Croatia (1994) (from 1952 to 1993 represented by Yugoslavia)*
Croatian Society of Mechanics

*Cyprus (2015) (from 2010 to 2014 Associate Organization)*
Cyprus Mathematical Society

*Czech Republic (1993) (from 1949 to 1992 represented by Czechaslowakia)*
The National Committee of Theoretical and Applied Mechanics

*Denmark* (1949)
National Committee for Theoretical & Applied Mechanics, The Royal Danish
Academy of Sciences and Letters

*Egypt* (1976)
Academy of Scientific Research and Technology, Egyptian Committee of
Theoretical and Applied Mechanics

*Estonia (1992) (from 1956 to 1991 represented by USSR)*
Estonian Committee for Mechanics

*Finland (1952)*
The Finnish National Committee on Mechanics

*France (1949)*
Comité National Français de Mécanique, Académie des Sciences

*Georgia (2000) (from 1956 to 1991 represented by USSR)*
National Committee of Theoretical and Applied Mechanics

*Germany (1950) (for some years (1973–1990) separately represented by Federal
Republic of Germany (FRG) and German Democratic Republic (GDR))*
Deutsches Komitee für Mechanik (DEKOMECH) within the Gesellschaft für
Angewandte Mathematik und Mechanik

*Greece (1979)*
Hellenic Society for Theoretical and Applied Mechanics

*Hungary (1948)*
Hungarian National Committee for IUTAM

*India (1950)*
The Indian National Committee for Theoretical and Applied Mechanics of the
Indian National Science Academy

*Ireland (1984)*
Irish National Committee for Theoretical and Applied Mechanics

*Israel (1950)*
The Israel Society for Theoretical and Applied Mechanics (ISTAM)

*Italy (1949)*
Associazione Italiana di Meccanica Teorica ed Applicata

*Japan (1951)*
The National Committee for Theoretical and Applied Mechanics of the Science Council of Japan

*Korea, Republic of (1990)*
Korean Committee for Theoretical and Applied Mechanics

*Latvia (1992–2014) (from 1956 to 1991 represented by USSR)*
Latvian National Committee for Mechanics

*Mexico (2008)*
Mexican Academy of Sciences

*Morocco (1998–2004)*
Societe Marocaine Des Sciences Mecaniques

*Netherlands (1952)*
Netherlands Mechanics Committee

*New Zealand (1979)*
The Royal Society of New Zealand, Committee on Mathematical & Information Sciences

*Norway (1949)*
National Committee on Theoretical and Applied Mechanics

*Poland (1952)*
Committee for Mechanics of the Polish Academy of Sciences

*Portugal (1968)*
Portuguese Society of Theoretical, Applied and Computational Mechanics

*Romania (1956)*
Romanian National Committee of Theoretical and Applied Mechanics

*Russia (1992) (from 1956 to 1991 represented by USSR)*
Russian National Committee on Theoretical and Applied Mechanics

*Saudi Arabia (1988)*
King Abdullaziz City for Science and Technology

*Serbia (2006) (from 1952 to 2002 represented by Yugoslavia, 2003–2005 representing Serbia and Montenegro)*
Serbian Society of Mechanics

*Slovakia (1993–2014) (from 1949 to 1992 represented by Czechaslowakia)*
The Slovak Society for Mechanics, Council of Scientific Societies

*Slovenia (1994) (from 1952 to 1993 represented by Yugoslavia)*
Slovene Mechanics Society

*South Africa (1994)*
National Research Foundation (NRF), South African Association for Theoretical
and Applied Mechanics (SAAM)

*Spain (1950)*
The National Institute of Aerospace Technology

*Sweden (1950)*
Swedish National Committee for Mechanics

*Switzerland (1950)*
Board of the Federal Institutes of Technology
(Rat der Eidgenössischen Technischen Hochschulen)

*Turkey (1950)*
Turkish National Committee of Theoretical and Applied Mechanics

*UK (1948)*
The Royal Society, UK Panel for IUTAM

*Ukraine (1995) (from 1956 to 1991 represented by USSR)*
National Committee of Ukraine on Theoretical and Applied Mechanics

*USA (1949)*
The U.S. National Committee on Theoretical and Applied Mechanics,

*Vietnam (1990)*
Vietnam Association of Mechanics (VAM)

# Appendix 20
# Affiliated Organizations

CISM (1970)
International Centre for Mechanical Sciences, Udine, Italy

ICHMT (1972)
International Centre for Heat and Mass Transfer

ICR (1974)
International Committee on Rheology

IAVSD (1978)
International Association for Vehicle System Dynamics

EUROMECH (1978)
European Mechanics Committee

ISIMM (1978)
International Society for the Interaction of Mechanics and Mathematics

ICF (1978)
International Congress on Fracture

ICM (1982)
International Congress on Mechanical Behavior of Materials

AFMC (1982)
Asian Fluid Mechanics Committee

IACM (1984)
International Association for Computational Mechanic

CACOFD (1992–2010, from then represented by LACCOTAM)
Caribbean Congress of Fluid Dynamics

IABEM (1994)
International Association for Boundary Element Methods

ISSMO (1996)
International Society for Structural and Multidisciplinary Optimization

© The Author(s) 2016
P. Eberhard and S. Juhasz (eds.), *IUTAM*,
DOI 10.1007/978-3-319-31063-3

HYDROMAG (1996)
International Association for Hydromagnetic Phenomena and Applications

IIAV (1997)
International Institute of Acoustics and Vibration

ICA (1998)
International Commission for Acoustics

ICTS (2002)
International Congresses on Thermal Stresses

BICTAM (2010)
Beijing International Center for Theoretical and Applied Mechanics

LACCOTAM (2010)
Latin American and Caribbean Conference on Theoretical and Applied Mechanics

IMSD (2014)
International Association of Multibody System Dynamics

IASCM (2014)
International Association for Structural Control and Monitoring

# Appendix 21
# Officers and Additional Bureau Members

| Year | President | Vice Pres. | Treasurer | Secretary |
|------|-----------|------------|-----------|-----------|
| 1948 | J. Pérès (France) | R.V. Southwell (UK) | H.L. Dryden (USA) | J.M. Burgers (Netherlands) |
| 1952 | H.L. Dryden (USA) | J. Pérès (France) | G. Temple (UK) | F.A.v.d. Dungen (Belgium) |
| 1956 | F.K.G. Odqvist (Sweden) | H.L. Dryden (USA) | G. Temple (UK) | M. Roy (France) |
| 1960 | G. Temple (UK) | F.K.G. Odqvist (Sweden) | W.T. Koiter (Netherlands) | M. Roy (France) |
| 1964 | M. Roy (France) | G. Temple (UK) | W.T. Koiter (Netherlands) | H. Görtler (Germany) |
| 1968 | W.T. Koiter (Netherlands) | M. Roy (France) | H. Görtler (Germany) | F.I. Niordson (Denmark) |
| 1972 | H. Görtler (Germany) | W.T. Koiter (Netherlands) | D.C. Drucker (USA) | F.I. Niordson (Denmark) |
| 1976 | F.I. Niordson (Denmark) | H. Görtler (Germany) | D.C. Drucker (USA) | J. Hult (Sweden) |
| 1980 | D.C. Drucker (USA) | F.I. Niordson (Denmark) | E. Becker (Germany) | J. Hult (Sweden) |
| 1984 | J. Lighthill (UK) | D.C. Drucker (USA) | L. van Wijngaarden (Netherlands) | W. Schiehlen (Germany) |
| 1988 | P. Germain (France) | Sir J. Lighthill (UK) | L. van Wijngaarden (Netherlands) | W. Schiehlen (Germany) |
| 1992 | L. van Wijngaarden (Netherlands) | P. Germain (France) | B.A. Boley (USA) | F. Ziegler (Austria) |
| 1996 | W. Schiehlen (Germany) | L. van Wijngaarden (Netherlands) | L.B. Freund (USA) | M.A. Hayes (Ireland) |
| 2000 | H.K. Moffatt (UK) | W. Schiehlen (Germany) | L.B. Freund (USA) | D.H. van Campen (Netherlands) |
| 2004 | L.B. Freund (USA) | H.K. Moffatt (UK) | J. Engelbrecht (Estonia) | D.H. van Campen (Netherlands) |
| 2008 | T.J. Pedley (UK) | L.B. Freund (USA) | N. Olhoff (Denmark) | F. Dias (France) |
| 2012 | V. Tvergaard (Denmark) | T.J. Pedley (UK) | P. Eberhard (Germany) | F. Dias (France) |

© The Author(s) 2016
P. Eberhard and S. Juhasz (eds.), *IUTAM*,
DOI 10.1007/978-3-319-31063-3

| Year | Member | Member | Member | Member |
|------|--------|--------|--------|--------|
| 1948 | F.A.v.d. Dungen (Belgium) | H. Solberg (Denmark) | H. Favre (Switzerland) | G. Colonetti (Italy) |
| 1952 | J.M. Burgers (Netherlands) | S. Goldstein (USA) | R. Grammel (Germany) | F.K.G. Odqvist (Sweden) |
| 1956 | J. Ackeret (Switzerland) | G. Colonetti (Italy) | R. Grammel (Germany) | W.T. Koiter (Netherlands) |
| 1960 | H. Görtler (Germany) | N.J. Hoff (USA) | N. Muskhelish-vili (USSR) | H. Ziegler (Switzerland) |
| 1964 | N.J. Hoff (USA) | W.Olszak (Poland) | H. Parkus (Austria) | L.I. Sedov (USSR) |
| 1968 | N.J. Hoff (USA) | J. Lighthill (UK) | W.Olszak (Poland) | L.I. Sedov (USSR) |
| 1972 | R. Legendre (France) | J. Lighthill (UK) | W.Olszak (Poland) | L.I. Sedov (USSR) |
| 1976 | P. Germain (France) | J. Lighthill (UK) | W.Olszak (Poland) | L.I. Sedov (USSR) |
| 1980 | P. Germain (France) | J. Lighthill (UK) | L.I. Sedov (USSR) | I. Tani (Japan) |
| 1984 | P. Germain (France) | J. Hult (Sweden) | A.Y. Ishlinsky (USSR) | I. Imai (Japan) |
| 1988 | B.A. Boley (USA) | G.G. Chernyi (USSR) | I. Imai (Japan) | F. Ziegler (Austria) |
| 1992 | G.G. Chernyi (USSR) | H.K. Moffatt (UK) | W. Schiehlen (Germany) | T. Tatsumi (Japan) |
| 1996 | J. Engelbrecht (Estonia) | H.K. Moffatt (UK) | T. Tatsumi (Japan) | Ren Wang (China) |
| 2000 | C. Cercignani (Italy) | J. Engelbrecht (Estonia) | R. Narasimha (India) | J. Salencon (France) |
| 2004 | T. Kambe (Japan) | A. Kluwick (Austria) | N. Olhoff (Denmark) | Z. Zheng (China) |
| 2008 | F. Chernousko (Russia) | I. Goldhirsch (Israel) | N.K. Gupta (India) | A. Thess (Germany) |
| 2012 | N. Aubry (USA) | M. Rubin (Israel) | B. Schrefler (Italy) | W. Yang (China) |

# Appendix 22
# List of Reports

All reports are available from the IUTAM web site www.iutam.org

| Year | Pages |
|------|-------|
| 1948 | 19 |
| 1949 | 20 |
| 1950 | 28 |
| 1951 | 36 |
| 1952 | 48 |
| 1953 | 46 |
| 1954,55,56 | 45 |
| 1957 | 37 |
| 1958 | 41 |
| 1959 | 36 |
| 1960 | 51 |
| 1961 | 45 |
| 1962 | 49 |
| 1963 | 44 |
| 1964 | 56 |
| 1965 | 60 |
| 1966 | 70 |
| 1967 | 61 |

| Year | Pages |
|------|-------|
| 1968 | 64 |
| 1969 | 70 |
| 1970 | 67 |
| 1971 | 66 |
| 1972 | 72 |
| 1973 | 66 |
| 1974 | 83 |
| 1975 | 76 |
| 1976 | 87 |
| 1977 | 80 |
| 1978 | 92 |
| 1979 | 91 |
| 1980 | 99 |
| 1981 | 79 |
| 1982 | 110 |
| 1983 | 88 |
| 1984 | 89 |
| 1985 | 85 |

| Year | Pages |
|------|-------|
| 1986 | 87 |
| 1987 | 91 |
| 1988 | 101 |
| 1989 | 96 |
| 1990 | 126 |
| 1991 | 109 |
| 1992 | 131 |
| 1993 | 146 |
| 1994 | 172 |
| 1995 | 188 |
| 1996 | 221 |
| 1997 | 183 |
| 1998 | 134 |
| 1999 | 129 |
| 2000 | 157 |

| Year | Pages |
|------|-------|
| 2001 | 140 |
| 2002 | 194 |
| 2003 | 161 |
| 2004 | 184 |
| 2005 | 164 |
| 2006 | 200 |
| 2007 | 182 |
| 2008 | 175 |
| 2009 | 145 |
| 2010 | 162 |
| 2011 | 151 |
| 2012 | 182 |
| 2013 | 135 |
| 2014 | 180 |
| | |

© The Author(s) 2016
P. Eberhard and S. Juhasz (eds.), *IUTAM*,
DOI 10.1007/978-3-319-31063-3

# IUTAM-ICTAM Photos

Group picture (Delft 1924)

P. Eberhard and S. Juhasz (eds.), *IUTAM*,
DOI 10.1007/978-3-319-31063-3

Group picture (Zurich 1926)

Group picture (Cambridge, UK 1934)

Group picture (Cambridge, USA 1938)

Group picture (London 1948)

T.v. Karman, C.B. Biezeno and others (Pallanza 1950, Symposium)

J.v. Neumann (Istanbul 1952)

Mrs. V. Hoff, Sir R. Southwell, N. Hoff and S. Gunturkin (Istanbul 1952)

L. Sedov, T.v. Karman, -?-, N. Brankov and N. Dolptchiev (Bruxelles 1956)

H.L. Dryden, P. Baudoux, F.v.d. Dungen, T.v. Karman, A.L. Jaumotte and
J. Pérès (Bruxelles 1956)

P. Germain (Bruxelles 1956)

T.v. Karman with Japanese delegates (Bruxelles 1956)

*Facing:* T.v. Karman, A.L. Jaumotte, J. Pérès, G.F. Temple, M.Roy,
N.I. Mushkelishvili and G. Colonetti.
*Forefront:* P. Germain, J.v.d. Kerkhove, and W. Prager (Bruxelles 1956)

N.I. Mushkelishvili, G. Colonetti, J.M. Burgers, -?-, R. Chan, L. Sedov and
G. Batchelor (Bruxelles 1956)

*First row:* Sir G. Taylor, J. Burgers, H.L. Dryden, J. Pérès, G. Colonetti.
*Second row:* F. Grammel, W. Tollmien, G.F. Temple, -?-and F.K.G. Odqvist
(Bruxelles 1956)

D. Drucker, P. Hodge and M. Biot (Bruxelles 1956)

J. Keller, -?-, E. Reissner, E. Carafoll, F. Kampe de Feriet, H. Schlichting,
W. Prager, R. Bishop, N. Goodier, S. Juhasz and A. Philips (Bruxelles 1956)

F. Kampé de Fériet, G. Batchelor, N. Hoff, M. Hetenyi and M.J. Lighthill
(Bruxelles 1956)

Soviet delegates and W. Nowacki (Bruxelles 1956)

J. Ackeret, J. Pérès, H.L. Dryden, T.v. Karman and G.F. Temple
(Bruxelles 1956)

L.I. Sedov, -?- and T.v. Karman (Stresa 1960)

C.S. Draper, -?-, J. Den Hartog, K.V. Frolov, S.H. Crandall and G.K. Mikhailov
(Kiev 1962, Symposium)

*First row:* W.T. Koiter, M. Roy, N.I. Muskhelishvili, F. Odqvist, E. Carafoli
*Second row:* G.K. Mikhailov, L Sneddon, L.N. Vekua and L.I. Sedov
(Tbilisi 1963, Symposium)

Congress Hall of the Deutsches Museum (Munich 1964)

L.I. Sedov, S. Timoshenko, E. Truckenbrodt, H. Görtler and H. Neuber
(Munich 1964)

I. Tani, N.I. Muskhelishvili, F.K.G. Odqvist, H. Görtler and W. Nash
(Munich 1964)

H. Wittmeier, E.B. Igenbergs and H. Boflev (Munich 1964)

B.R. Seth and W. Olszak (Munich 1964)

E.B. Igenbergs and L.I. Sedov (Munich 1964)

W. Olszak, R. Sauer, Mrs Parkus, -?-, Sir R. Southwell and F. Rimrott
(Munich 1964)

N.J. Hoff opens Congress (Stanford 1968)

W. Prager and Mrs Prager (Stanford 1968)

Mrs. V. Hoff, O. Ljungstrom, Sir G. Taylor and N. Goodier (Stanford 1968)

C.R. Steele with some USSR delegates (Stanford 1968)

Congress Place in the Kremlin, site of opening ceremony (Moscow 1972)

G.K. Mikhailov at Congress opening (Moscow 1972)

G. Mikhailov, H.N. Abramson, W.T. Koiter, F. Niordson and S. Juhasz
(Moscow 1972)

Past Famous Mechanics Scientists Exhibit (Moscow 1972)

P. Francis, A.d. Pater, Rabotnov, S. Juhasz, W. Prager, F. Niordson, J. Besseling,
G. Batchelor, N. Luikov, F. Odqvist, L.I. Sedov and W. Fiszdon
(Delft 1976)

From Delft To Delft-Exhibit (Delft 1976)

Opening session (Toronto 1980)

F. Niordson, Lt. Governor of Ontario, F. Rimrott and -?-, -?-  (Toronto 1980)

B. Boley, B. Tabarrok, D. Drucker and A. Ishlinsky at Hart House Reception
(Toronto 1980)

H. Leutheusser, J. Hult, K. Charbonneau, F. Rimrott,
D.C. Drucker and S. Juhasz (Toronto 1980)

Opening Session – Ancient Horns (See Logo) (Lyngby 1984)

N. Olhoff, Mrs A. Drucker, D.C. Drucker, -?-, -?- and F. Niordson
(Lyngby 1984)

Mrs. A. Drucker, D.C. Drucker, F. Niordson, -?-, -?- and N. Olhoff
(Lyngby 1984)

IUTAM/ICTAM - A Short History – Exhibit (Lyngby 1984)

Bureau Meeting (Stuttgart 1987)
*from right:* D. Drucker, J. Hult, A. Ishlinsky, J. Lighthill, W. Schiehlen,
L.v. Wijngaarden, P. Germain, I. Imai

F. Ziegler (Stuttgart 1989, Symposium)      E. Becker (Toronto 1980)

J. Lighthill giving his speech, on his right and left sides are P. Germain and
M. Piau (Grenoble 1988)

The Opening Ceremony, on the podium from right to left:
A. Solan, L.v. Wijngaarden, S.R. Bodner, J. Lighthill, P. Germain,
Mayor A. Gurel, J. Singer, H.K. Moffatt, Z. Hashin, W. Schiehlen (Haifa 1992)

W. Schiehlen, K. Gersten, I. Imai Symposia Panel (Haifa 1992)

J. Lighthill, W. Schiehlen (Haifa 1992)

T. Tatsumi, Opening Ceremony (Kyoto 1996)

J. Lighthill (Kyoto 1996)

L.v. Wijngaarden (Kyoto 1996)

Executive Committee of the Congress Committee. H. Aref, S. Bodner, T. Pedley,
A. Acrivos, W. Schiehlen, N. Olhoff, R. Moreau (Aalborg 1999)

Bureau 1996-2000 Meeting (Chicago 2000)
W. Schiehlen, K. Moffatt, L.v. Wijngaarden, B. Freund, M. Hayes,
R. Wang, J. Engelbrecht, T. Tatsumi,
J. Salençon (2000-2004), D. van Campen (2000-2004)

Welcome Reception (Chicago 2000)
*from left:* J. Olhoff, F. Niordson, ICTAM 2000 President H. Aref,
N. Alhoff and S. Aref

*W. Schiehlen, N. Olhoff,  J. Phillips, H. Aref (Chicago 2000)*

R. Christensen and B. Freund (Chicago 2000)

Closing Lecture K. Moffatt (Chicago 2000)

Closing Ceremony W. Schiehlen (Chicago 2000)

W. Schiehlen, S. Crandall (Cambridge 2002, General Assembly)

Bureau Meeting (Cambridge 2002)
D.v. Campen, J. Salençon, J. Engelbrecht, R. Narasimha,
C. Cercignani, K. Moffatt, W. Schiehlen, B. Freund

Opening Lecture L.v. Wijngaarden (Warsaw 2004)

Bureau Meeting (Providence 2006)
*from left:* A. Kluwick, T. Kambe, K. Moffatt, Z. Zheng, B. Freund,
J. Engelbrecht, D. van Campen, N. Olhoff

F. Dias, D. van Campen (Paris 2010, General Assembly)

P. Eberhard, D. van Campen (Paris 2010, General Assembly)

A. van Campen-Stuurman, W. & C. Schiehlen (Beijing 2012)

P. Eberhard, G. Ambrósio, J. Ambrósio (Beijing 2012)

A. Fidlin, F. Pfeiffer, P. Eberhard (Beijing 2012)

D. van Campen, W. Yang (Beijing 2012)

Executive Committee of the Congress Committee (Stuttgart 2015).
*from left:* B. Eckhardt, V. Tvergaard, M. Floryan (President ICTAM Montreal),
D. van Campen, B. McMeeking, G. Stepan, J. Magnaudet

Bureau Meeting (Stuttgart 2015)
*from left:* M. Rubin, B. Schrefler, P. Eberhard,
V. Tvergaard, F. Dias, N. Aubry, T. Pedley

Printed in the United States
By Bookmasters